A2-Level
Physics

The Revision Guide

Editor:
Julie Wakeling

Contributors:
Tony Alldridge, Abraham Baravi, Stuart Barker, Jane Cartwright, Peter Cecil, Peter Clarke, Mark A.Edwards, Sarah Hilton, D. Kamya, Barbara Mascetti, John Myers, Sam Norman, Zoe Nye, Alan Rix, Alice Shepperson, Moira Steven, Sharon Watson, Andy Williams, Tony Winzor.

Proofreaders:
Ian Francis, Zoe Nye, Steve Parkinson, Glenn Rogers.

Published by Coordination Group Publications Ltd.

This book covers all the core modules for:
Edexcel, AQA A, AQA B, OCR A and OCR B.

It also covers the optional modules:
AQA A Module 5 — Astrophysics
AQA A Module 6 — Medical Physics
AQA A Module 8 — Turning Points in Physics
OCR A Option 01 — Cosmology
OCR A Option 02 — Health Physics
OCR A Option 04 — Nuclear and Particle Physics

There are notes at the tops of pages to tell you whether or not you need it for your syllabus.

Many thanks to the HST programme at CERN and the University of Birmingham for their kind permission to reproduce the photographs used on p104 & 105.

ISBN: 1-84146-367-1
Groovy website: www.cgpbooks.co.uk
Jolly bits of clipart from CorelDRAW
Printed by Elanders Hindson, Newcastle upon Tyne.

Contents

Momentum and Impulse

*These pages are for **AQA B, OCR A** and **OCR B** only.*

This section is about linear momentum — that's momentum in a straight line (not a circle).

Understanding **Momentum** helps you do **Calculations** on **Collisions**

The **momentum** of an object depends on two things — **mass** and **velocity**.
The **product** of these two values is the momentum of the object.

Remember, velocity is a vector quantity, so it has size <u>and</u> direction.

momentum = mass × velocity	or in symbols:	p (in kg ms^{-1}) = m (in kg) × v (in ms^{-1})

Momentum is always **Conserved**

1)   Assuming **no external forces** act, momentum is always **conserved**.

2)   This means the **total momentum** of two objects **before** they collide **equals** the total momentum **after** the collision.

3)   This is really handy for working out the **velocity** of objects after a collision (as you do...):

Example   A skater of mass 75 kg and velocity 4 ms^{-1} collides with a stationary skater of mass 50 kg. The two skaters join together and move off in the same direction. Calculate their velocity after impact.

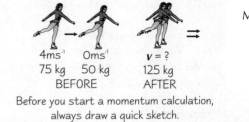

4ms^{-1} 0ms^{-1} v = ?
75 kg   50 kg   125 kg
BEFORE   AFTER

Before you start a momentum calculation, always draw a quick sketch.

Momentum of skaters before = momentum of skaters after
$$(75 \times 4) + (50 \times 0) = 125\,v$$
$$300 = 125\,v$$
$$\text{So } v = 2.4 \text{ ms}^{-1}$$

4)   The same principle can be applied in **explosions**. E.g. if you fire an **air rifle**, the **forward momentum** gained by the pellet **equals** the **backward momentum** of the rifle, and you feel the rifle recoiling into your shoulder.

Example   A bullet of mass 0.005 kg is shot from a rifle at a speed of 200 ms^{-1}. The rifle has a mass of 4 kg. Calculate the velocity at which the rifle recoils.

4 kg x v 0.005 kg x 200 ms^{-1}

Momentum before explosion = Momentum after explosion
$$0 = (0.005 \times 200) + (4 \times v)$$
$$0 = 1 + 4v$$
$$v = -0.25 \text{ ms}^{-1}$$

Rocket Propulsion can be Explained by Momentum

For a **rocket** to be **propelled forward** it must expel **exhaust gases**. The momentum of the rocket in the forward direction is **equal** to the momentum of the exhaust gases in the backward direction.

Example   A rocket of mass 500 kg is completely stationary in an area of space a long way from any gravitational fields. It starts its engines. The rocket ejects 2.0 kg of gas per second at a speed of 1000 ms^{-1}. Calculate the velocity of the rocket after 1 second.
(For the purpose of this calculation ignore the loss of mass due to fuel use.)

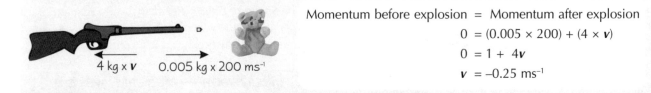

2 kg ←   500 kg
1000 ms^{-1} v = ?

The total momentum before rocket fires = 0 kg ms^{-1} as initially the rocket is stationary; the total momentum after rocket fires must also = 0 kg ms^{-1}.

Total momentum after rocket fires = $(500 \times v) + (2 \times 1000) = 0$

$$500\,v = -2000$$
$$v = -4 \text{ ms}^{-1}$$

The minus sign shows that the rocket moves in the opposite direction to the exhaust gases.

Momentum and Impulse

Collisions can be Elastic or Inelastic
OCR A only   See page 9 for more on this.

An **elastic collision** is one where **momentum** is **conserved** and **kinetic energy** is **conserved** — i.e. no energy is dissipated as heat, sound, etc. If a collision is **inelastic** it means that some of the kinetic energy is converted into other forms during the collision. But **momentum is always conserved.**

Example
A toy lorry (mass 2 kg) travelling at 3 ms⁻¹ crashes into a smaller toy car (mass 800 g), travelling in the same direction at 2 ms⁻¹. The velocity of the lorry after the collision is 2.6 ms⁻¹ in the same direction. Calculate the new velocity of the car and the total kinetic energy before and after the collision.

2kg
3ms⁻¹
BEFORE

800g
2ms⁻¹

v = 2.6ms⁻¹ v = ?
AFTER

Momentum before collision = Momentum after collision

$(2 \times 3) + (0.8 \times 2)$ = $(2 \times 2.6) + (0.8\, v)$

7.6 = $5.2 + 0.8\, v$

2.4 = $0.8\, v$

v = 3 ms^{-1}

Kinetic Energy before = k.e. of lorry + k.e. of car

$= \tfrac{1}{2}mv^2 \text{ (lorry)} + \tfrac{1}{2}mv^2 \text{ (car)}$

$= \tfrac{1}{2}(2 \times 3^2) + \tfrac{1}{2}(0.8 \times 2^2)$

$= 9 + 1.6$

$= 10.6 \text{ J}$

Kinetic Energy after = $\tfrac{1}{2}(2 \times 2.6^2) + \tfrac{1}{2}(0.8 \times 3^2)$

$= 6.76 + 3.6$

$= 10.36 \text{ J}$

The difference in the two values (10.6 J – 10.36 J = 0.24 J) is the amount of kinetic energy <u>dissipated</u> as heat or sound, or in damaging the vehicles — so this is an <u>inelastic collision</u>.

Impulse = Change in Momentum
AQA B only

1) Newton's Second Law says **force = rate of change of momentum** (see pages 4-5), or $F = (mv - mu) \div t$

2) **Rearranging** Newton's 2ⁿᵈ Law gives: ⟹
Where **impulse** is defined as **force × time**, *Ft*.
The units of impulse are **newton seconds**, Ns.

> $Ft = mv - mu$
> (where *v* is the final velocity and *u* is the initial velocity)
> so *impulse = change of momentum*

3) So, the **force** of an impact can be **reduced** by **increasing the time** of the impact.

For example, a toy car with a mass of 1 kg, travelling at 5 ms⁻¹, hits a wall and stops in a time of 0.5 seconds.

The force on the car is: $F = \dfrac{mv - mu}{t} = \dfrac{(1 \times 5) - (1 \times 0)}{0.5} = 10 \text{ N}$

But if the time of impact is doubled to 1 second, the force on the car is halved.

This is the idea behind car crumple zones which increase the time of an impact to reduce the force on the passengers.

Practice Questions

Q1 Give two examples of conservation of momentum in practice.

Q2 Describe what happens when a tiny object makes an elastic collision with a massive object, and why.

Exam Questions

Q1 A ball of mass 0.6 kg moving at 5 ms⁻¹ collides with a larger stationary ball of mass 2 kg.
The smaller ball rebounds in the opposite direction at 2.4 ms⁻¹. What is the velocity of the larger ball? [3 marks]

Q2 A toy train of mass 0.7 kg, travelling at 0.3 ms⁻¹, collides with a stationary toy carriage of mass 0.4 kg.
The two toys couple together. What is their new velocity? [3 marks]

This ~~ain't~~ is rocket science...

*It seems a bit of a contradiction to say that momentum's always conserved then tell you that impulse is the change in momentum. The difference is that impulse is only talking about the change of momentum of one of the objects, whereas conservation of momentum is talking about the **whole** system.*

Newton's Laws of Motion

These pages are for OCR A and OCR B only.

You did most of this at GCSE, but that doesn't mean you can just skip over it now. You'll be kicking yourself if you forget this stuff in the exam — easy marks...

Newton's 1st Law says that a Force is Needed to Change Velocity

1) **Newton's 1st law of motion** states the **velocity** of an object will **not change** unless a **resultant force** acts on it.

2) In plain English this means a body will remain at rest or moving in a **straight line** at a **constant speed**, unless acted on by a **resultant force.**

An apple sitting on a table won't go anywhere because the **forces** on it are **balanced.**

reaction (R)	=	**weight** (mg)
(force of table pushing apple up)		(force of gravity pulling apple down)

3) If the forces **aren't balanced**, the **overall resultant force** will cause the body to **accelerate.** This may involve a change in **direction**, or **speed**, or both. (See Newton's 2nd law, below.)

Newton's 2nd Law says that Force is the Rate of Change in Momentum...

...which can be written as the well known equation:

resultant force (F) = mass (m) × acceleration (a)

Learn this — it crops up all over the place in A2 Physics. And learn what it means too:

1) It says that the **more force** you have acting on a certain mass, the **more acceleration** you get.

2) It says that for a given force the **more mass** you have, the **less acceleration** you get.

REMEMBER:
1) The **resultant force** is the **vector sum** of all the forces.
2) The force is **always** measured in **newtons. Always.**
3) The **mass** is always measured in **kilograms.**
4) a is the **acceleration** of the object as a result of **F.** It's **always** measured in **metres per second per second** (ms^{-2}).
5) The **acceleration** is always in the **same direction** as the **resultant force.**

Newton didn't just write down the equation **F = ma**, what he discovered was this:

> "The **rate of change of momentum** of an object is **directly proportional** to the **resultant force** which acts on the object."

so $F = \dfrac{\Delta mv}{\Delta t}$

Galileo said: All Objects Fall at the Same Rate (if you Ignore Air Resistance)

You need to understand **why** this is true. Newton's 2nd law explains it neatly — consider two balls dropped at the same time — ball **1** being heavy, and ball **2** being light. Then use Newton's 2nd law to find their acceleration.

mass = m_1 resultant force = F_1 acceleration = a_1 By Newton's Second Law: $\qquad F_1 = m_1 a_1$ Ignoring air resistance, the only force acting on the ball is weight, given by $W_1 = m_1 g$ (where g = gravitational field strength = 9.81 Nkg^{-1}). So: $F_1 = m_1 a_1 = W_1 = m_1 g$ So: $m_1 a_1 = m_1 g$, then m_1 cancels out to give: $a_1 = g$	mass = m_2 resultant force = F_2 acceleration = a_2 By Newton's Second Law: $\qquad F_2 = m_2 a_2$ Ignoring air resistance, the only force acting on the ball is weight, given by $W_2 = m_2 g$ (where g = gravitational field strength = 9.81 Nkg^{-1}). So: $F_2 = m_2 a_2 = W_2 = m_2 g$ So: $m_2 a_2 = m_2 g$, then m_2 cancels out to give: $a_2 = g$

... in other words, the **acceleration** is **independent of the mass.** It makes **no difference** whether the ball is **heavy or light**. And I've kindly **hammered home the point** by showing you two almost identical examples.

Newton's Laws of Motion

Newton's 3rd Law of Motion:

Each Force has an **Equal**, **Opposite Reaction Force**

There's a few different ways of stating Newton's 3rd law, but the clearest way is:

> **If an object A EXERTS a FORCE on object B, then**
> **object B exerts AN EQUAL BUT OPPOSITE FORCE on object A**

You'll also hear the law as "every action has an equal and opposite reaction".
But this confuses people who wrongly think the forces are both applied to the same object.
(If that were the case, you'd get a resultant force of zero and nothing would ever move anywhere...)
So remember — the two forces **act on different objects**.

The best way to get it is by looking at loads of examples:

1) If you **push against a wall**, the wall will **push back** against you, **just as hard**. As soon as you stop pushing, so does the wall. Amazing...

2) If you think about it, there must be an **opposing force** when you lean against a wall — otherwise you (and the wall) would **fall over**.

3) If you **pull a cart**, whatever force **you exert** on the rope, the rope exerts the **exact opposite** pull on you (unless the rope's stretching).

4) When you **sit** on a large **ill-tempered mutated** goose, your weight exerts a **downwards** force on the goose, while you feel the **equal force** of the goose **pushing upwards** on you.

Newton's 3rd law applies in **all situations** and to all **types of force**. But the pairs of forces are always the **same type**, e.g. both gravitational or both electrical.

This looks like Newton's 3rd law...

But it's NOT.

Gravity pulls down on book

Table pushes upwards on book

...because both forces are acting on the book.
The forces are equal and opposite, resulting in zero acceleration, so this is showing Newton's 1st law.

Practice Questions

Q1   State Newton's 1st, 2nd and 3rd laws of motion, and explain what they mean.

Q2   What are the pair of forces acting between an orbiting satellite and the Earth?

Exam Questions

Q1   Draw diagrams to show the forces acting on a parachutist:
   (i)   at the moment of jumping from a plane.   [1 mark]
   (ii) accelerating downwards.   [1 mark]
   (iii) having reached terminal velocity.   [1 mark]

Q2   A boat is moving across a river. The engines provide a force of 500 N at right angles to the flow of the river and the boat experiences a drag of 100 N in the opposite direction. The force on the boat due to the flow of the river is 300 N. The mass of the boat is 250 kg.
   (a) Calculate the magnitude of the resultant force acting on the boat.   [2 marks]
   (b) Calculate the magnitude of the acceleration of the boat.   [2 marks]

Newton's three incredibly important laws of motion...

These equations may not really fill you with a huge amount of excitement (and I hardly blame you if they don't)... but it was pretty fantastic at the time — suddenly people actually understood how forces work, and how they affect motion. I mean arguably it was one of the most important scientific discoveries ever...

Work and Power

These pages are for AQA B, OCR A and OCR B only.

As everyone knows, work in Physics isn't like normal work. It's harder. Work also has a specific meaning that's to do with movement and forces. You'll have seen this at GCSE — it just comes up in more detail for A2.

Work is done whenever Energy is Transferred

This table gives you some examples of **work being done** and the **energy changes** that happen.

1) Usually you need a force to move something because you're having to **overcome another force**.

2) The thing being moved has **kinetic energy** while it's **moving**.

3) The kinetic energy is transferred to **another form of energy** when the movement stops.

ACTIVITY	WORK DONE AGAINST	FINAL ENERGY FORM
Lifting up a box.	gravity	gravitational potential energy
Pushing a chair across a level floor.	friction	heat (thermal)
Pushing two magnetic north poles together.	magnetic force	magnetic energy
Stretching a spring.	stiffness of spring	elastic potential energy

The word **'work'** in Physics means the **amount of energy transferred** from one form to another when a force causes a movement of some sort.

Work = Force × Distance

When a car tows a caravan, it applies a force to the caravan and moves it to where you want it to go. To **find out** how much **work** has been **done**, you need to use the **equation**:

> **work done (W) = force causing motion (F) × distance moved (s)**
>
> ...where **W** is measured in joules (J), **F** is measured in newtons (N) and **s** is measured in metres (m).

Points to remember:

1) **Work** is the **energy** that's been **changed** from one form to another — it's not necessarily the **total** energy. E.g. moving a book from a low shelf to a higher one will increase its gravitational potential energy, but it had some potential energy to start with. Here, the **work done** would be the **increase** in potential energy, **not the total** potential energy.

2) Remember, the distance needs to be measured in metres — if you have **distance in centimetres or kilometres**, you need to **convert** to metres first.

3) The force **F** will be a **fixed** value in any calculations, either because it's **constant** or because it's the **average** force.

4) The equation assumes that the **direction of the force** is the **same** as the **direction of movement**.

5) The equation gives you the **definition** of the joule (symbol J): 'one joule is the work done when a force of 1 newton moves an object through a distance of 1 metre'

6) If you plotted a graph of force (**F**) against distance moved (**s**), the **area under the graph** would equal the work done.

The Force isn't always in the Same Direction as the Movement

Sometimes the **direction of movement** is **different** from the **direction of the force**.

Example

1) To **calculate the work done** in a situation like the one in the diagram, you need to consider the **horizontal** and **vertical components** of the **force**.

2) The only **movement** is in the **horizontal** direction. This means the **vertical force** is not causing any motion (and hence not doing any work) — it's just **balancing** out some of the **weight**, meaning there's a **smaller reaction force**.

direction of force on sledge

rosebud

direction of motion

3) The horizontal force is causing the motion — so to **calculate the work done**, this is the **only force** you need to consider. Which means we get:

> $W = Fs\cos\theta$

Where θ is the **angle** between the **direction of the force** and the **direction of motion**.

F / θ / $F\cos\theta$ → Direction of motion

Work and Power

Power = Work Done per Second

Power means many things in everyday speech, but in physics (of course!) it has a special meaning. Power is the **rate of doing work** — in other words it is the **amount of energy transformed** from one form to another **per second.**
You **calculate power** with this equation:

> **Power (P) = work done (W) / time (t)**
> ...where P is measured in watts (W), W is measured in joules (J) and t is measured in seconds (s)

The **watt** (symbol W) is defined as a **rate of energy transfer** equal to **1 joule per second**.
Yep, that's another **equation and definition** for you to **learn**.

Power is also Force × Velocity (P = Fv)

Sometimes, it's **easier** to use **this version** of the power equation. This is how you get it:

1)   You **know** $P = W/t$
2)   You also **know** $W = Fs$, which gives $P = Fs/t$ (if the object is moving at a constant velocity).
3)   But $v = s/t$, which you can substitute into the above equation to give $P = Fv$
4)   It's easier to use this if you're given the **speed** in the question.
        Learn this equation as a **shortcut** to link **power** and **speed**.

Example

A car is travelling at a speed of $10 \, \text{ms}^{-1}$ which is kept going against the frictional force by a driving force of $500 \, \text{N}$ in the direction of motion. Find the power supplied by the engine to keep the car moving.

Use the shortcut $P = Fv$, which gives:
$P = 500 \times 10 = 5000 \, \text{W}$

If the force and motion are in different directions, you can replace F with $F\cos\theta$ (where θ is the **angle** between the **direction of the force** and the **direction of motion**) to get:

$$P = Fv\cos\theta$$

You **aren't** expected to **remember** this equation, but it's made up of bits that you **are supposed to know**, so be ready for the possibility of calculating **power** in a situation where the **direction of the force and direction of motion are different**.

Practice Questions

Q1   Write down the equation used to calculate work if the force and motion are in the same direction.

Q2   Write down the equation for work if the force is at an angle to the direction of motion.

Q3   Write down the equations relating (i) power and work, and (ii) power and speed.

Exam Questions

Q1   A traditional narrowboat is drawn by a horse walking along a towpath.
        The horse pulls the boat at a constant speed between two locks which are
        1500 m apart. The tension in the rope is 100 N at 40° to the direction of motion.

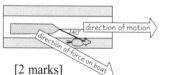

(a) How much work is done on the boat?   [2 marks]
(b) The boat moves at $0.8 \, \text{ms}^{-1}$. Calculate the power supplied to the boat in the direction of motion.   [2 marks]

Q2   A motor is used to lift a 20 kg load a height of 3 m. (Take g = 9.81 Nkg^{-1}.)
(a) Calculate the work done in lifting the load.   [2 marks]
(b) The speed of the load during the lift is $0.25 \, \text{ms}^{-1}$. Calculate the power delivered by the motor.   [2 marks]

Work, work, work — when will it all end..*

So work is the amount of energy needed for a force to move something a certain distance, and power is the rate that work is done — easy. Now all you need to do is learn the equations and you'll be fine...

* Answer: page 163

Conservation of Energy

These pages are for AQA B, OCR A and OCR B only.

Energy can never be **lost**. *I repeat — **energy** can **never** be lost. Which is basically what I'm about to take up two whole pages saying. But that's, of course, because you need to do exam questions on this as well as understand the principles.*

Learn the **Principle** of **Conservation** of **Energy**

The **principle of conservation of energy** says that:

Energy **cannot be created** or **destroyed**. Energy **can be transferred** from one form to another but the total amount of energy in a closed system will not change.

Example

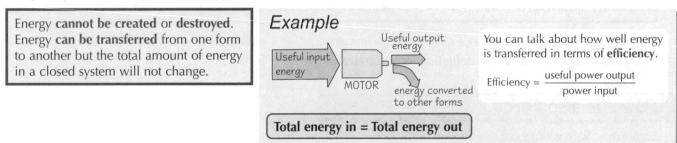

Useful input energy → MOTOR → Useful output energy / energy converted to other forms

You can talk about how well energy is transferred in terms of **efficiency**.

$$\text{Efficiency} = \frac{\text{useful power output}}{\text{power input}}$$

Total energy in = Total energy out

You need it for **Questions** about **Kinetic** and **Potential Energy**

The Principle of Conservation of Energy nearly always comes up when you're doing questions about changes between kinetic and potential energy.

A quick reminder:

1) **Kinetic energy** is energy of anything **moving** which you work out from $E_k = \frac{1}{2}mv^2$, where v is the velocity it's travelling at and m is its mass.

2) There are **different types of potential energy** — e.g. gravitational and elastic.

3) **Gravitational potential energy** is the energy something gains if you lift it up. You work it out using: $\Delta E_p = mg\Delta h$, where m is the mass of the object, Δh is the height it is lifted and g is the gravitational field strength ($9.81\,\text{Nkg}^{-1}$ on Earth).

4) **Elastic potential energy** (elastic stored energy) is the energy you get in, say, a stretched rubber band or spring. You work this out using $E = \frac{1}{2}ke^2$, where e is the extension of the spring and k is the stiffness constant.

Examples   These pictures show you three **examples** of changes between kinetic and potential energy.

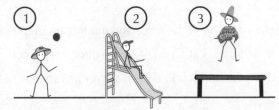

1) As Becky throws the **ball upwards**, **kinetic energy** is converted into **gravitational potential energy**. When it **comes down** again, that **gravitational potential** energy is **converted back** into **kinetic** energy.

2) As Dominic goes **down the slide**, **gravitational potential energy** is converted to **kinetic energy**.

3) As Simon bounces upwards from the trampoline, **elastic potential energy** is converted to **kinetic energy**, to **gravitational potential energy**. As he comes back down again, that **gravitational potential** energy is **converted back** to **kinetic** energy, to **elastic potential** energy, and so on.

In **real life** there are also **frictional forces** — Simon would have to use some **force** from his **muscles** to keep **jumping** to the **same height** above the trampoline each time. Each time the trampoline **stretches**, some **heat** is generated in the trampoline material. You're usually told to **ignore friction** in exam questions — this means you can **assume** that the **only forces** are those that provide the **potential energy** (in this example that's **Simon's weight** and the **tension** in the springs and trampoline material).

If you're ignoring friction, you can say that the **sum of the kinetic and potential energies is constant**.

Conservation of Energy

In a Collision, Kinetic Energy is Conserved — but Only if it's Elastic

OCR A only

1) As long as there's **no friction**, you know that **momentum is always conserved** in a collision (you have the **same total momentum after** a collision **as you had before**) (see pages 2 and 3).

2) **After** a collision, objects sometimes **stick together**, and sometimes **bounce apart**. Either way the momentum is still **conserved**.

3) But the **kinetic energy** is **not** always conserved. Usually, some of it gets converted into **sound or heat** energy.

In the real world, some energy's always lost in a collision. Sometimes, if the energy loss is small, it's okay to assume the collision is elastic.

A collision where the **total kinetic energy** is the **same** after a collision is called an **elastic collision**.

A collision where the **total kinetic energy** is **less** after a collision is called an **inelastic collision**.

In the diagram on the right, given no friction,

1) it is **always true** that $m_1v_1 + m_2v_2 = m_1v_3 + m_2v_4$

2) it is **sometimes true** that $\frac{1}{2}m_1v_1^2 + \frac{1}{2}m_2v_2^2 = \frac{1}{2}m_1v_3^2 + \frac{1}{2}m_2v_4^2$

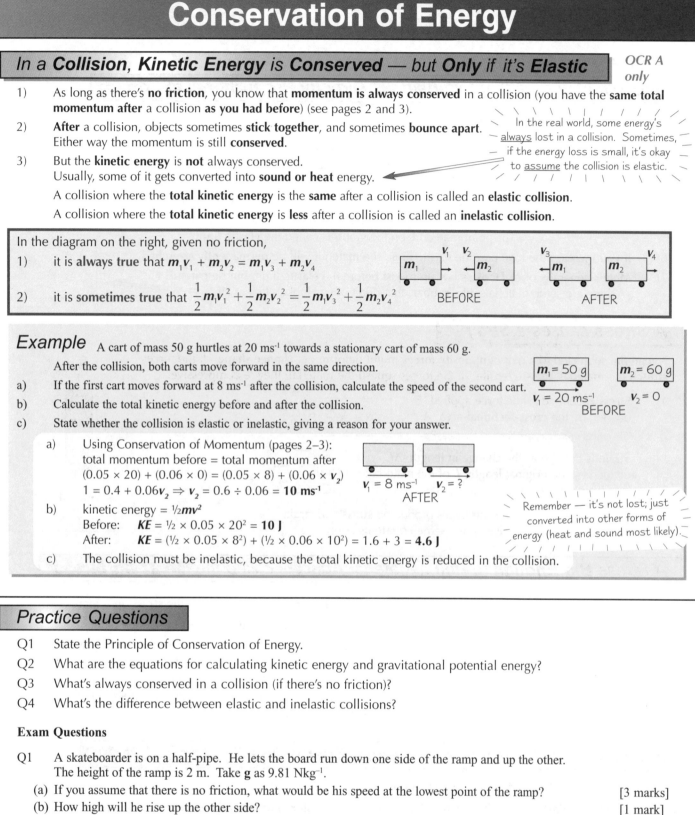

Example

A cart of mass 50 g hurtles at 20 ms⁻¹ towards a stationary cart of mass 60 g. After the collision, both carts move forward in the same direction.

a) If the first cart moves forward at 8 ms⁻¹ after the collision, calculate the speed of the second cart.

b) Calculate the total kinetic energy before and after the collision.

c) State whether the collision is elastic or inelastic, giving a reason for your answer.

a) Using Conservation of Momentum (pages 2–3):
total momentum before = total momentum after
$(0.05 \times 20) + (0.06 \times 0) = (0.05 \times 8) + (0.06 \times v_2)$
$1 = 0.4 + 0.06v_2 \Rightarrow v_2 = 0.6 \div 0.06 = \textbf{10 ms}^{-1}$

b) kinetic energy = $\frac{1}{2}mv^2$
Before: $KE = \frac{1}{2} \times 0.05 \times 20^2 = \textbf{10 J}$
After: $KE = (\frac{1}{2} \times 0.05 \times 8^2) + (\frac{1}{2} \times 0.06 \times 10^2) = 1.6 + 3 = \textbf{4.6 J}$

Remember — it's not lost; just converted into other forms of energy (heat and sound most likely).

c) The collision must be inelastic, because the total kinetic energy is reduced in the collision.

Practice Questions

Q1 State the Principle of Conservation of Energy.

Q2 What are the equations for calculating kinetic energy and gravitational potential energy?

Q3 What's always conserved in a collision (if there's no friction)?

Q4 What's the difference between elastic and inelastic collisions?

Exam Questions

Q1 A skateboarder is on a half-pipe. He lets the board run down one side of the ramp and up the other. The height of the ramp is 2 m. Take **g** as 9.81 Nkg⁻¹.
(a) If you assume that there is no friction, what would be his speed at the lowest point of the ramp? [3 marks]
(b) How high will he rise up the other side? [1 mark]
(c) Real ramps are not frictionless, so what must the skater do to reach the top on the other side? [1 mark]

Q2 A railway truck of mass 10 000 kg is travelling at 1 ms⁻¹ and collides with a stationary truck of mass 15 000 kg. The two trucks stay together after the collision.
(a) What can you say about the total kinetic energy before and after the collision without calculating anything? [1 mark]
(b) Calculate the final velocity of the two trucks. [2 marks]
(c) Calculate the total kinetic energy before and after the collision. [2 marks]

Conserve your energy — you're going to need it for Quantum Physics...

As my Gran always tells me, 'What you put in is what you get out' — this is a universal truth that works equally well for energy, physics revision and beer. So I guess there's no arguing with me Gran (well, her umbrella always deterred me...).

Stretching Solids

These pages are for AQA B only.

Hooke's Law says that Extension is Directly Proportional to Force

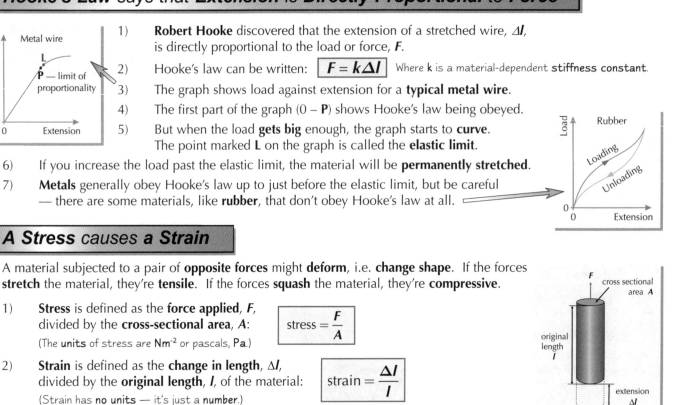

1) **Robert Hooke** discovered that the extension of a stretched wire, Δl, is directly proportional to the load or force, **F**.

2) Hooke's law can be written: $\boxed{F = k\Delta l}$ *Where* k *is a material-dependent* **stiffness constant**.

3) The graph shows load against extension for a **typical metal wire**.

4) The first part of the graph (0 – **P**) shows Hooke's law being obeyed.

5) But when the load **gets big** enough, the graph starts to **curve**. The point marked **L** on the graph is called the **elastic limit**.

6) If you increase the load past the elastic limit, the material will be **permanently stretched**.

7) **Metals** generally obey Hooke's law up to just before the elastic limit, but be careful — there are some materials, like **rubber**, that don't obey Hooke's law at all.

A Stress causes a Strain

A material subjected to a pair of **opposite forces** might **deform**, i.e. **change shape**. If the forces **stretch** the material, they're **tensile**. If the forces **squash** the material, they're **compressive**.

1) **Stress** is defined as the **force applied**, *F*, divided by the **cross-sectional area**, *A*:

$$\text{stress} = \frac{F}{A}$$

(The **units** of stress are Nm^{-2} or pascals, **Pa**.)

2) **Strain** is defined as the **change in length**, Δl, divided by the **original length**, *l*, of the material:

$$\text{strain} = \frac{\Delta l}{l}$$

(Strain has **no units** — it's just a **number**.)

3) It doesn't matter whether the forces producing **stress** and **strain** are **tensile** or **compressive** — the **same equations** apply.

A Stress Big Enough to Pull a Material Apart is called the Breaking Stress

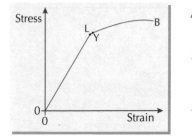

As a greater and greater **force** is applied to a material, the **stress** on it **increases**.

1) The effect of the **stress** is to **pull** the **atoms apart** from each other.

2) Eventually the stress becomes **so great** that atoms **separate completely**, and the **material breaks**. This is shown by point **B** on the graph. The stress at which this occurs is called the **breaking stress**.

3) The point marked **Y** on the graph is called the **yielding stress**. At this point the material suddenly 'gives'.

Elastic Stored Energy is the Potential Energy in a Stretched Material

When a material is **stretched**, **work** has to be done in stretching the material.

1) **Before** the **elastic limit**, all the **work done** in stretching is **stored** as **potential energy** in the material.

2) This stored energy is called **elastic strain energy**.

3) On a **graph** of force against extension, the elastic strain energy is given by the **area under the graph**.

Provided a material obeys Hooke's law, the **potential energy** stored inside it can be **calculated** quite easily.

1) The work done on the wire in stretching it is equal to the energy stored.

2) **Work done** equals **force × displacement**.

3) However, the **force** on the material **isn't constant**. It rises from zero up to force **F**. To calculate the **work done**, use the average force between zero and **F**, i.e. ½**F**.

4) The work done is equal to the **elastic stored energy**, E_s, so: $\boxed{E_s = \tfrac{1}{2}F\Delta l}$

Stretching Solids

The *Young Modulus* is Stress ÷ Strain

When you **stretch** a material, it experiences a **tensile stress** and a **tensile strain**.

1) Up to a point called the **limit of proportionality**, the stress and strain of a material are directly proportional to each other.

2) So, for a particular material, stress divided by strain is a constant, called the **Young modulus**, *E*.

$$E = \frac{\text{tensile stress}}{\text{tensile strain}} = \frac{F/A}{\Delta l/l} = \frac{Fl}{A\Delta l}$$

Where, *F* = force in N, *A* = cross-sectional area in m², *l* = initial length in m and *Δl* = extension in m.

3) The **units** of the Young modulus are the same as stress (**Nm⁻²** or pascals), since strain has no units.

4) If you plot a graph of stress against strain, the gradient of the graph gives the Young modulus.

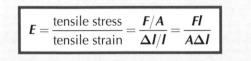

$E = \dfrac{\text{stress}}{\text{strain}} = \text{gradient}$

Stress / Strain graph

Practice Questions

Q1 Define Hooke's law.

Q2 Define tensile forces and compressive forces.

Q3 Explain what is meant by the elastic limit of a material.

Q4 What is the difference between stress and strain? Write a definition for each.

Q5 How can the elastic strain energy be found from the force against extension graph of a stretched material?

Q6 Define the Young modulus for a material.

Exam Questions

Q1 A metal guitar string stretches 4.0 mm when a 10 N force is applied.
  (a) If the string obeys Hooke's law, how far will the string stretch with a 15 N force? [1 mark]
  (b) Calculate the stiffness constant for this string in Nm⁻¹. [2 marks]
  (c) The string is tightened beyond its elastic limit. What would be noticed about the string? [1 mark]

Q2 A rubber band is 6.0 cm long. When it is loaded with 2.5 N, its length becomes 10.4 cm. Further loading increases the length to 16.2 cm when the force is 5.0 N.
  Does the rubber band obey Hooke's law? Justify your answer with a suitable calculation. [2 marks]

Q3 A steel wire is 2.00 m long. When a 300 N force is applied to the wire, it stretches 4.0 mm. The wire has a circular cross-section with a diameter of 1.0 mm.
  (a) Calculate the tensile stress in the wire. [2 marks]
  (b) Calculate the tensile strain of the wire. [1 mark]
  (c) What is the value of the elastic stored energy in the stretched wire? [1 mark]
  (d) What is the value of the Young modulus for steel? [1 mark]

Hey amigo — don't stress, it's gonna be alright...

A few more equations to learn, but apart from that it's all fairly straightforward stuff here — so no excuses for forgetting anything on these pages. Right then, if you've learnt it all, I reckon you deserve a nice cup of tea and a double chocolate chip muffin... then come back for a little more mind-bending physics revision (no pain, no gain, no pain, no gain, no pain...)

Circular Motion

These pages are for Edexcel, AQA A, AQA B, OCR A and OCR B.

*It's probably worth putting a bookmark in here — this stuff is needed **all over** the place.*

Angles can be Expressed in Radians

The angle in **radians**, θ, is defined as the **arc-length** divided by the radius of the circle.

For a **complete circle** (360°), the arc-length is just the circumference of the circle ($2\pi r$). Dividing this by the radius (r) gives 2π. So there are 2π radians in a complete circle.

$$\text{angle in radians} = \frac{2\pi}{360} \times \text{angle in degrees}$$

1 radian is about 57°

Some common angles:

45°
$\frac{\pi}{4}$ rad

90°
$\frac{\pi}{2}$ rad

180°
π rad

The Angular Speed is the Angle an Object Rotates Through per Second

1) Just as **linear speed**, v, is defined as distance ÷ time, the **angular speed**, ω, is defined as **angle ÷ time**. The unit is rad s⁻¹ — radians per second.

$$\omega = \frac{\theta}{t}$$

ω = angular speed (rad s⁻¹) — the symbol for angular speed is the little Greek 'omega', not a w.
θ = angle (radians) turned through in a time, t (seconds)

2) The **linear speed**, v, and **angular speed**, ω, of a rotating object are linked by the equation:

$$v = r\omega$$

v = linear speed (ms⁻¹), r = radius of the circle (m), ω = angular speed (rad s⁻¹)

Example — Beam of Particles in a Cyclotron   (See Section 9)

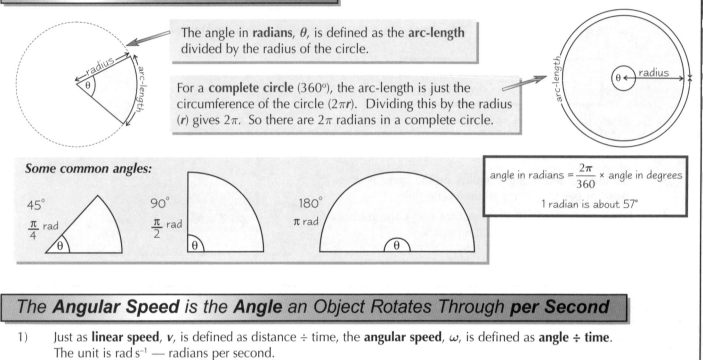

FAST

SLOW

All parts of the beam take the same time to rotate through this angle.

1) Different parts of the particle beam are rotating at **different linear speeds**, v. (The linear speed is sometimes called **tangential velocity**.)

2) But all the parts **rotate** through the **same angle** in the **same time** — so they have the same **angular speed**.

Circular Motion has a Frequency and Period

1) The frequency, f, is the number of complete **revolutions per second** (rev s⁻¹ or hertz, Hz).

2) The period, T, is the **time taken** for a complete revolution (in seconds).

3) Frequency and period are **linked** by the equation:

$$f = \frac{1}{T}$$

f = frequency in rev s⁻¹, T = period in s

4) For a complete circle, an object turns through 2π radians in a time T, so frequency and period are related to ω by:

$$\omega = 2\pi f \quad \text{and} \quad \omega = \frac{2\pi}{T}$$

f = frequency in rev s⁻¹, T = period in s, ω = angular speed in rad s⁻¹

Circular Motion

Objects Travelling in Circles are **Accelerating** since their **Velocity is Changing**

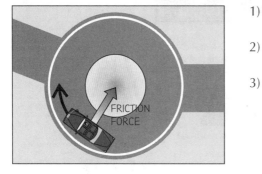

1) Even if the car shown is going at a **constant speed**, its **velocity** is changing since its **direction** is changing.

2) Since acceleration is defined as the **rate of change of velocity**, the car is accelerating even though it isn't going any faster.

3) This acceleration is called the **centripetal acceleration** and is always directed towards the **centre of the circle**.

$$a = \frac{v^2}{r} \quad \text{and} \quad a = \omega^2 r$$

a = centripetal acceleration in ms⁻²
v = linear speed in ms⁻¹
ω = angular speed in rad s⁻¹
r = radius in m

The **Centripetal Acceleration** is produced by a **Centripetal Force**

From Newton's laws, if there's a **centripetal acceleration**, there must be a **centripetal force** acting towards the **centre of the circle**. Since $F = ma$, the centripetal force must be:

$$F = \frac{mv^2}{r} \quad \text{and} \quad F = m\omega^2 r$$

The object will fly off at a tangent if the centripetal force is removed.

Men cowered from the force of the centipede.

You Feel **Weightless** when you're in **Freefall**   *Edexcel only*

1) If you're in a lift and the cable breaks, you feel **weightless**, because you're hurtling towards the ground with the **same acceleration as the lift** (i.e. 9.81 ms⁻²). If you jump off the floor you'll 'float' up to the ceiling.

2) The same thing happens when you're in **orbit** (apart from the hurtling to a grisly death bit). Your **centripetal acceleration** towards the centre to the Earth is the **same** as the **spacecraft's**.

3) You **feel** weightless even though gravity is still acting on you.

Practice Questions

Q1 How many radians are there in a complete circle?
Q2 How is angular speed defined and what is the relationship between angular speed and linear speed?
Q3 Define the period and frequency of circular motion. What is the relationship between period and angular speed?
Q4 In which direction does the centripetal force act, and what happens when this force is removed?

Exam Questions

Q1 (a) At what angular speed does the Earth orbit the Sun? (1 year = 3.2 × 10⁷ s)   [2 marks]
    (b) Calculate the Earth's linear speed. (Assume radius of orbit = 1.5 × 10¹¹ m)   [2 marks]
    (c) Calculate the centripetal force needed to keep the Earth in its orbit. (Mass of Earth = 6.0 × 10²⁴ kg)   [2 marks]
    (d) What is providing this force?   [1 mark]

Q2 A bucket full of water, tied to a rope, is being swung around in a vertical circle (so it is upside down at the top of the swing).
    (a) By considering the acceleration due to gravity at the top of the swing, what is the minimum frequency with which the bucket can be swung without any water falling out? (radius of circle = 1 m)   [3 marks]
    (b) The bucket is now swung with a constant angular speed of 5 rad s⁻¹. What will be the tension in the rope when the bucket is at the top of the swing if the total mass of the bucket and water = 10 kg?   [2 marks]

I'm spinnin' around, move out of my way...

*"Centripetal" just means "centre-seeking". The centripetal force is what that actually causes circular motion. What you **feel** when you're spinning, though, is the reaction (centrifugal) force. Don't get the two mixed up.*

Simple Harmonic Motion

These pages are for Edexcel, AQA A, AQA B, OCR A and OCR B.

SHM is Defined in terms of Acceleration and Displacement

1) An object moving with **simple harmonic motion** (SHM) **oscillates** to and fro, either side of a **midpoint**.

2) The distance of the object from the midpoint is called its **displacement**.

3) There is always a **restoring force** pulling or pushing the object back **towards** the **midpoint**.

4) The **size** of the **restoring force** depends on the **displacement**, and the force makes the object **accelerate** towards the midpoint:

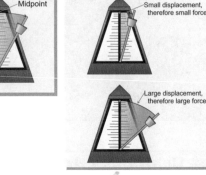

> **SHM:** an oscillation in which the **acceleration** of an object is **directly proportional** to its **displacement** from the midpoint, and is directed **towards the midpoint**.

The Restoring Force makes the Object Exchange PE and KE

1) The **type** of **potential energy** (PE) depends on **what it is** that's providing the **restoring force**. This will be **gravitational PE** for pendulums and **elastic PE** (elastic stored energy) for masses on springs.

2) As the object moves **towards the midpoint**, the restoring force **does work** on the object and so **transfers** some **PE** to **KE**. When the object is moving **away from the midpoint**, all that KE is transferred **back to PE** again.

3) At the **midpoint**, the object's **PE** is **zero** and its **KE** is **maximum**.

4) At the **maximum displacement** (the **amplitude**) on both sides of the midpoint, the object's **KE** is **zero** and its **PE** is **maximum**.

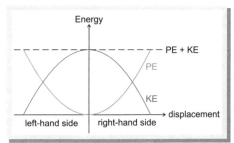

5) The **sum** of the **potential** and **kinetic** energy is called the **mechanical energy** and **stays constant** (as long as the motion isn't damped — see p. 18).

6) The **energy transfer** for one complete cycle of oscillation (see graph) is: PE to KE to PE to KE to PE … and then the process repeats…

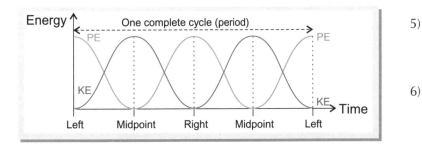

You can Draw Graphs to Show Displacement, Velocity and Acceleration

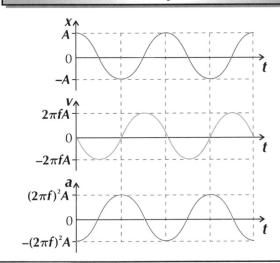

Displacement, **x**, varies as a cosine or sine wave with a maximum value, **A** (the amplitude) or x_0 if you're doing Edexcel.

Velocity, **v**, is the gradient of the displacement-time graph (dx/dt). It has a maximum value of $(2\pi f)A$ (where f is the frequency of the oscillation) and is a quarter of a cycle in front of the displacement.

Acceleration, **a**, is the gradient of the velocity-time graph (d^2x/dt^2). It has a maximum value of $(2\pi f)^2A$, and is in antiphase with the displacement.

Simple Harmonic Motion

The Frequency and Period don't depend on the Amplitude

1) From **maximum positive displacement** (e.g. maximum displacement to the left) to **maximum negative displacement** (e.g. maximum displacement to the right) and **back again** is called a **cycle** of oscillation.

2) The **frequency**, *f*, of the SHM is the number of cycles per second (measured in Hz).

3) The **period**, *T*, is the **time** taken for a complete cycle (in seconds).

> In SHM, **frequency** and **period** are independent of the **amplitude** (i.e. constant for a given oscillation). So a pendulum clock will keep ticking in regular time intervals even if its swing becomes very small.

Learn the SHM Equations

You all need to know the equations in the green boxes.

1) According to the definition of SHM, the **acceleration**, *a* (or d²x/dt² — **OCR B**), is directly proportional to the **displacement**, *x*.
The **constant of proportionality** depends on the **frequency**. The **minus sign** is there because the acceleration is always in the **opposite direction** from the displacement.

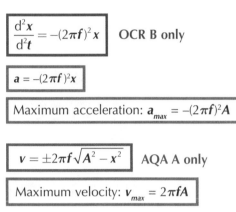

$$\frac{d^2x}{d^2t} = -(2\pi f)^2 x$$ OCR B only

$$a = -(2\pi f)^2 x$$

Maximum acceleration: $a_{max} = -(2\pi f)^2 A$

2) The **velocity** is **positive** if the object's moving **away** from the **midpoint**, and **negative** if it's moving **towards** the midpoint. Hence the ± sign in the velocity equation.

$$v = \pm 2\pi f \sqrt{A^2 - x^2}$$ AQA A only

Maximum velocity: $v_{max} = 2\pi f A$

3) The **displacement** varies with time according to two equations depending on **where** the object was when the timing was started.

For someone starting a stopwatch with a pendulum at **maximum displacement**:

$$x = A\cos(2\pi ft)$$

For someone releasing a pendulum but starting a stopwatch as the pendulum swings through **the midpoint**:

$$x = A\sin(2\pi ft)$$

Practice Questions

Q1 Sketch a graph of how the velocity of an object oscillating with SHM varies with time.

Q2 What is the special relationship between the acceleration and the displacement in SHM?

Q3 Given the amplitude and the frequency, how would you work out the maximum acceleration?

Exam Questions

Q1 (a) Define *simple harmonic motion*. [2 marks]

(b) Explain why the motion of a ball bouncing off the ground is not SHM. [1 mark]

Q2 A pendulum is pulled a distance 0.05 m from its midpoint and released.
It oscillates with simple harmonic motion with a frequency of 1.5 Hz. Calculate:

(a) Its maximum velocity [1 mark]

(b) Its displacement 0.1 s after it is released [2 marks]

(c) The time it took to fall to 0.01 m from the midpoint after it is released [2 marks]

"Simple" harmonic motion — hmmm, I'm not convinced...

The basic concept of SHM is simple enough (no pun intended). Make sure you can remember the shapes of all the graphs on page 14 and the equations from this page, then just get as much practice at using the equations as you can.

Simple Harmonic Oscillators

This page is for Edexcel, AQA A, AQA B and OCR B only.

A *Mass* on a *Spring* is a *Simple Harmonic Oscillator (SHO)*

1) When the mass is **pushed above** or **pulled below** the **equilibrium position**, there's a **force** exerted on it.

2) The size of this force is:

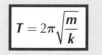
$$F = -kx$$
where **k** is the **spring constant** (stiffness) of the spring in Nm⁻¹ (see page 11) and **x** is the displacement in m.

3) After a bit of jiggery-pokery involving Newton's second law and some of the ideas on the previous page, you get the **formula for the period of a mass oscillating on a spring**:

$$T = 2\pi\sqrt{\frac{m}{k}}$$

where **T** = period of oscillation in seconds
m = mass in kg
k = spring constant in Nm⁻¹

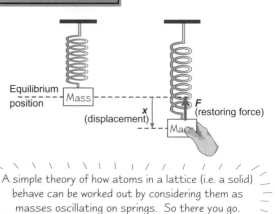

Equilibrium position — Mass
x (displacement)
F (restoring force)

A simple theory of how atoms in a lattice (i.e. a solid) behave can be worked out by considering them as masses oscillating on springs. So there you go.

You can check this result **EXPERIMENTALLY** by changing **one variable at a time** and seeing what happens.

Investigating the Mass-Spring System

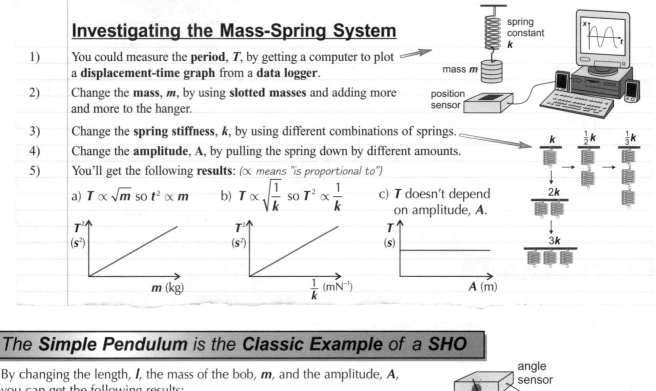

1) You could measure the **period, T**, by getting a computer to plot a **displacement-time graph** from a **data logger**.

2) Change the **mass, m**, by using **slotted masses** and adding more and more to the hanger.

3) Change the **spring stiffness, k**, by using different combinations of springs.

4) Change the **amplitude, A**, by pulling the spring down by different amounts.

5) You'll get the following **results**: (∝ means "is proportional to")

a) $T \propto \sqrt{m}$ so $t^2 \propto m$

b) $T \propto \sqrt{\frac{1}{k}}$ so $T^2 \propto \frac{1}{k}$

c) T doesn't depend on amplitude, **A**.

The *Simple Pendulum* is the *Classic Example* of a *SHO*

By changing the length, **l**, the mass of the bob, **m**, and the amplitude, **A**, you can get the following results:

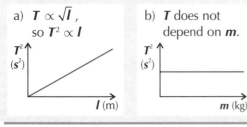

a) $T \propto \sqrt{l}$, so $T^2 \propto l$

b) T does not depend on **m**.

c) T does not depend on **A**.

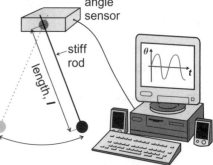

The **formula for the period of a pendulum** is: (The derivation's quite hard, so you don't need to know it.)

This formula only works for small angles of oscillation — up to about 10° from the equilibrium point.

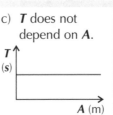
$$T = 2\pi\sqrt{\frac{l}{g}}$$

where **T** = period of oscillation in seconds
l = length of pendulum (between pivot and centre of mass of bob) in m
g = gravitational field strength in Nkg⁻¹

Simple Harmonic Oscillators

This page is for AQA B only.

Using a Pendulum to Measure g

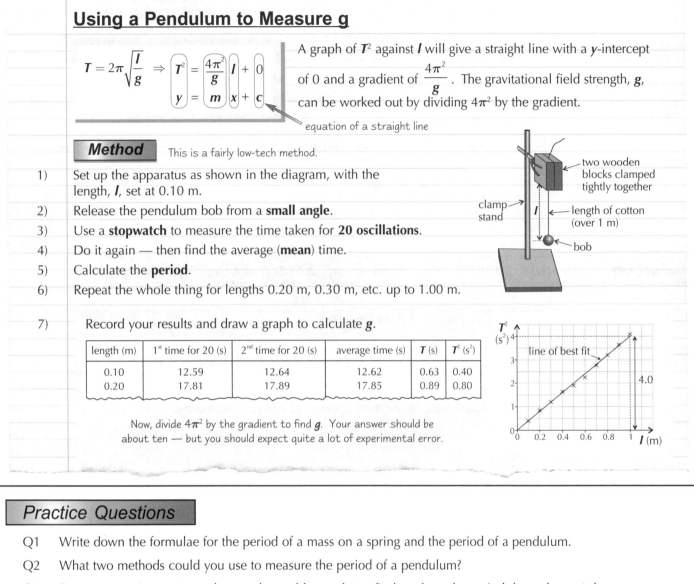

$$T = 2\pi\sqrt{\frac{l}{g}} \Rightarrow \underset{y}{T^2} = \underset{m}{\left(\frac{4\pi^2}{g}\right)}\underset{x}{l} + \underset{c}{0}$$

equation of a straight line

A graph of T^2 against l will give a straight line with a y-intercept of 0 and a gradient of $\frac{4\pi^2}{g}$. The gravitational field strength, g, can be worked out by dividing $4\pi^2$ by the gradient.

Method   This is a fairly low-tech method.

1) Set up the apparatus as shown in the diagram, with the length, l, set at 0.10 m.
2) Release the pendulum bob from a **small angle**.
3) Use a **stopwatch** to measure the time taken for **20 oscillations**.
4) Do it again — then find the average (**mean**) time.
5) Calculate the **period**.
6) Repeat the whole thing for lengths 0.20 m, 0.30 m, etc. up to 1.00 m.
7) Record your results and draw a graph to calculate **g**.

length (m)	1st time for 20 (s)	2nd time for 20 (s)	average time (s)	T (s)	T² (s²)
0.10	12.59	12.64	12.62	0.63	0.40
0.20	17.81	17.89	17.85	0.89	0.80

Now, divide $4\pi^2$ by the gradient to find **g**. Your answer should be about ten — but you should expect quite a lot of experimental error.

Practice Questions

Q1   Write down the formulae for the period of a mass on a spring and the period of a pendulum.

Q2   What two methods could you use to measure the period of a pendulum?

Q3   For a mass-spring system, what graphs could you plot to find out how the period depends on a) the mass, b) the spring constant and c) the amplitude of a mass oscillating on a spring? What would they look like?

Exam Questions

Q1   A spring of original length 0.10 m is suspended from a stand and clamp.
A mass of 0.10 kg is attached to the bottom and the spring extends to a total length of 0.20 m.

(a) Calculate the spring constant of the spring in Nm⁻¹. ($g = 9.81$ Nkg⁻¹) See p. 11 for help with this one.   [2 marks]

(b) The mass is pulled down a further 2 cm and then released.
Calculate the period of the subsequent oscillations.   [1 mark]

(c) What mass would be needed to make the period of oscillation twice as long?   [2 marks]

Q2   Two pendulums of different lengths were released from rest at the top of their swing.
It was found that it took exactly the same time for the shorter pendulum to make five complete oscillations as it took the longer pendulum to make three complete oscillations.
If the shorter pendulum had a length of 0.20 m, what was the length of the longer one?   [3 marks]

Go on — SHO the examiners what you're made of...

The most important things to remember on these pages are those two period equations. You'll be given them in your exam, but you need to know what they mean and be happy using them.

Free and Forced Vibrations

These pages are for Edexcel, AQA A, AQA B, OCR A and OCR B.

Resonance… hmm… tricky little beast. Remember the Millennium Bridge, that standard-bearer of British engineering? The wibbles and wobbles were caused by resonance. How was it sorted out? By damping, which is on the next page.

Free Vibrations — No Transfer of Energy To or From the Surroundings

1) If you stretch and release a mass on a spring, it oscillates at its **natural frequency**.

2) If **no energy's transferred** to or from the surroundings, it will **keep** oscillating with the **same amplitude forever**.

3) In practice this **never happens**, but a spring vibrating in air is called a **free vibration** anyway.

> *OCR B only*: you need to know this formula for the **total energy** of a freely oscillating mass on a spring:
>
> $$E_{total} = \frac{1}{2}mv^2 + \frac{1}{2}kx^2 \text{ (in other words, K.E. + P.E.)}$$

> *AQA B only*: the **total energy** of a freely oscillating mass on a spring is **proportional** to the **amplitude squared**.

Forced Vibrations *happen when there's an* External Driving Force

1) A system can be **forced** to vibrate by a periodic **external force**.

2) The frequency of this force is called the **driving frequency**.

Resonance *happens when* Driving Frequency = Natural Frequency

When **driving frequency = natural frequency**, the system gains more and more energy from the driving force and so vibrates with a **rapidly increasing amplitude**. When this happens the system is **resonating**.

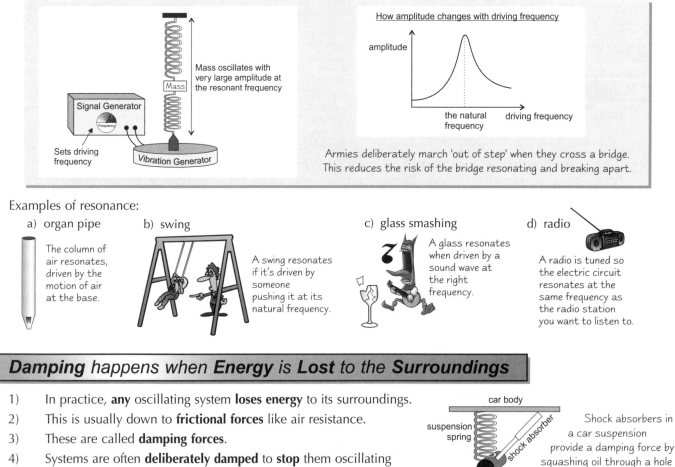

Armies deliberately march 'out of step' when they cross a bridge. This reduces the risk of the bridge resonating and breaking apart.

Examples of resonance:

a) organ pipe — The column of air resonates, driven by the motion of air at the base.

b) swing — A swing resonates if it's driven by someone pushing it at its natural frequency.

c) glass smashing — A glass resonates when driven by a sound wave at the right frequency.

d) radio — A radio is tuned so the electric circuit resonates at the same frequency as the radio station you want to listen to.

Damping *happens when* Energy *is* Lost *to the* Surroundings

1) In practice, **any** oscillating system **loses energy** to its surroundings.

2) This is usually down to **frictional forces** like air resistance.

3) These are called **damping forces**.

4) Systems are often **deliberately damped** to **stop** them oscillating or to **minimise** the effect of **resonance**.

Shock absorbers in a car suspension provide a damping force by squashing oil through a hole when compressed.

Free and Forced Vibrations

Different Amounts of Damping have Different Effects

1) The **degree** of damping can vary from **light** damping (where the damping force is small) to **overdamping**.

2) Damping **reduces** the **amplitude** of the oscillation over time. The **heavier** the damping, the **quicker** the amplitude is reduced to zero.

3) **Critical damping** reduces the amplitude (i.e. stops the system oscillating) in the **shortest possible time**.

4) Car **suspension systems** and moving coil **meters** are critically damped so that they **don't oscillate** but return to equilibrium as quickly as possible.

5) Systems with **even heavier damping** are **overdamped**. They take **longer** to return to equilibrium than a critically damped system.

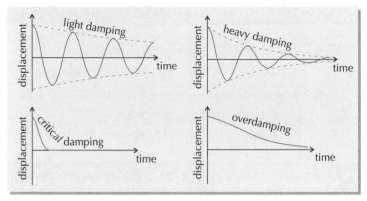

Damping Affects Resonance too

1) **Lightly damped** systems have a **very sharp** resonance peak. Their amplitude only increases dramatically when the **driving frequency** is **very close** to the **natural frequency**.

2) **Heavily damped** systems have a **flatter response**. Their amplitude doesn't increase very much and they aren't as **sensitive** to the driving frequency.

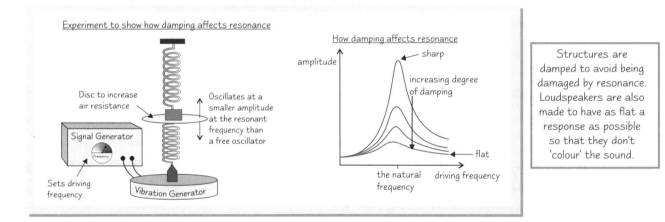

Structures are damped to avoid being damaged by resonance. Loudspeakers are also made to have as flat a response as possible so that they don't 'colour' the sound.

Practice Questions

Q1 What is a free vibration? What is a forced vibration?

Q2 Draw diagrams to show how a damped system oscillates with time when the system is lightly damped and when the system is critically damped.

Exam Questions

Q1 (a) What is resonance? [2 marks]

   (b) Draw a diagram to show how the amplitude of a lightly damped system varies with driving frequency. [2 marks]

   (c) On the same diagram, show how the amplitude of the system varies with driving frequency when it is heavily damped. [1 mark]

Q2 (a) What is critical damping? [1 mark]

   (b) State a situation where critical damping is used. [1 mark]

A2 Physics — it can really put a damper on your social life...

Resonance can be really useful (radios, oboes, swings — yay) or very, very bad...

The Nature of Waves

These pages are for Edexcel and AQA A only.

Aaaah... playing with slinky springs and waggling ropes about. It's all good clean fun as my mate Richard used to say...

A **Wave Transfers Energy** Away from its Source

A **progressive** (moving) wave carries **energy** from one place to another **without transferring any material**.
Here are some ways you can tell waves carry energy:

1) Electromagnetic waves cause things to **heat up**.

2) **X-rays** and **gamma rays** knock electrons out of their orbits, causing **ionisation**.

3) Loud **sounds** make things **vibrate**.

4) **Wave power** can be used to **generate electricity**.

Here are all the **Bits** of a **Wave** you Need to Know

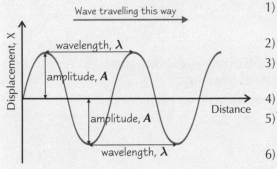

1) **Displacement**, *X*, metres — how far a **point** on the wave has **moved** from its **undisturbed position**.

2) **Amplitude**, *A*, metres — **maximum displacement**.

3) **Wavelength**, λ, metres — the **length** of **one whole wave**, from **crest** to **crest** or **trough** to **trough**.

4) **Period**, *T*, seconds — the **time taken** for a **whole vibration**.

5) **Frequency**, *f*, hertz — the **number** of **vibrations per second** passing a given **point**.

6) **Phase difference** — the amount by which **one wave lags behind another** wave. **Measured** in **degrees** or **radians**. See page 24.

The frequency is the **inverse** of the period, so:

$$Frequency = \frac{1}{period}$$

Make sure you get the **units** straight: **1 Hz = 1 s⁻¹**.

All waves display **reflection**, **refraction**, **diffraction** (see p. 28) and **interference** (see p. 30).

Wave Speed, Frequency and Wavelength are Linked by the Wave Equation

Wave speed can be measured just like the speed of anything else:

$$\text{Speed } (v) = \frac{\text{distance moved } (d)}{\text{time taken } (t)}$$

Remember, you're not measuring how fast a physical point (like one molecule of rope) moves. You're measuring how fast a point on the **wave pattern** moves.

Learn the **Wave Equation**...

Speed of wave (*v*) = frequency (*f*) × wavelength (λ)

$$v = f\lambda$$

You need to be able to rearrange this equation for f or λ.

In **Longitudinal Waves** the **Vibrations** are **Along** the Direction of Travel

The most **common** example of a **longitudinal wave** is **sound**. A sound wave consists of alternate **compressions** and **rarefactions** of the **medium** it's travelling through. (That's why sound can't go through a vacuum.) Some types of **earthquake shock waves** are also longitudinal.

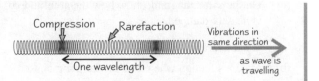

It's hard to **represent** longitudinal waves **graphically**. You'll usually see them plotted as **displacement** against **time**. These graphs can be **confusing**, though, because they look like a **transverse wave** (see next page).

The Nature of Waves

In *Transverse Waves* the *Vibration* is at *Right Angles* to the *Direction* of *Travel*

All **electromagnetic waves** are **transverse**.
Other examples of transverse waves are **ripples**
on water and waves on **ropes**.

Vibrations from side to side
Wave travelling this way

There are Two Main Ways of Showing Transverse Waves on Paper

Transverse waves can
be shown as **graphs**
of **displacement**
against **distance**
travelled by the wave.

crest
λ
Distance
trough
Displacement

They can also
be shown as
graphs of
displacement
against time.

P
Time
Displacement

Both sorts of graph often give the **same shape**, so pay attention to the label on the *x*-axis.
Displacements **upwards** from the centre line are given a **+ sign**. Displacements **downwards** are given a **– sign**.

A *Polarised Transverse Wave* only *Oscillates* In One Direction

1) If you **shake a rope** to make a **wave**, you can move your hand **up and down** or **side to side** or in a **mixture** of directions — it still makes a **transverse wave**.

2) But if you try to pass **waves in a rope** through a **vertical fence**, the wave will only get through if the **vibrations** are **vertical**. The fence filters out vibration in other directions. This is called **polarising** the wave.

direction of waves
rope
fence

3) Ordinary **light waves** are a mixture of **different directions** of **vibration**. (The things vibrating are electric and magnetic fields.) A **polarising filter** only transmits vibrations in one direction.

4) If you have two polarising filters at **right angles** to each other, then **no** light will get through.

5) Polarisation **can only happen** for **transverse** waves. The fact that you can polarise light is one **proof** that it's a transverse wave.

Practice Questions

Q1 Give the units of frequency, displacement and amplitude.
Q2 Write down the equation connecting *v*, λ and *f*.
Q3 Does a wave carry matter **or** energy from one place to another?
Q4 Diffraction and interference are two wave properties. Write down two more.
Q5 Write down two pieces of evidence that waves carry energy.
Q6 Explain what polarising sunglasses do to light.

Exam Questions

Q1 A buoy floating on the sea takes 6 seconds to rise and fall once (complete a full period of oscillation).
The difference in height between the buoy at its lowest and highest points is 1.2 m, and waves pass it at a speed of 3 ms^{-1}.

(a) How long are the waves? [2 marks]

(b) What is the amplitude of the waves? [1 mark]

Ooo, ooo, ooooooo good vibrations — bop, bop...

There's nothing particularly ground-shaking here — just a load of basic definitions and a couple of formulas to learn.
But don't even think about missing this stuff out. If you don't know it, the rest of the section won't make a lot of sense.

Electromagnetic Waves and Intensity

This page is for Edexcel and AQA A Option 8 only.

There's nothing really deep and meaningful to understand on this page — just a load of facts to learn I'm afraid.

Electromagnetic Waves are Transverse

1) They are **transverse** waves consisting of **vibrating electric** and **magnetic fields** at **right angles** to each other and the **direction of travel**. Electromagnetic waves are produced by **moving charged particles**.

2) All electromagnetic waves travel in a **vacuum** at a **speed** of **2.998 × 10⁸ ms⁻¹**, and at slower speeds in other media.

> **Just for AQA A Option 8 — Turning Points in Physics**
>
> In 1862, **James Clerk Maxwell** calculated the **speed of light** in a vacuum using:
>
> $$c = \frac{1}{\sqrt{\mu_0 \varepsilon_0}}$$
>
> where c is the speed of the wave in ms⁻¹, μ_0 ("mu-nought") is the permeability of free space (a constant — $4\pi \times 10^{-7}$ Hm⁻¹) and ε_0 ("epsilon-nought") is the permittivity of free space (another constant — 8.85×10^{-12} Fm⁻¹).

3) Like all waves, EM waves can be **reflected**, **refracted**, **diffracted**, undergo **interference** and obey $v = f\lambda$ (v = velocity, f = frequency, λ = wavelength).

4) Like all progressive waves, progressive EM waves **carry energy**.

5) Like all transverse waves, EM waves can be **polarised**.

> In 1887, **Heinrich Hertz** produced and detected **radio waves** using electric sparks. He showed by **experiment** that they could be reflected, refracted, diffracted and polarised, and show interference.

Some Properties Vary Across the EM Spectrum

EM waves with different wavelengths behave differently in some respects. The spectrum is split into seven categories: **radio waves**, **microwaves**, **infrared**, **visible light**, **ultraviolet**, **X-rays** and **gamma rays**.

1) The longer the wavelength, the more **obvious** the wave characteristics — e.g., long radio waves diffract round hills.

2) **Energy** is directly proportional to **frequency**. **Gamma rays** have the **highest energy**; **radio waves** the **lowest**.

3) The **higher** the **energy**, in general the more **dangerous** the wave.

4) The **lower the energy** of an EM wave, the **further from the nucleus** it comes from. **Gamma radiation** comes from inside the **nucleus**. **X-rays to visible light** come from energy-level transitions in **atoms** (see p. 66). **Infrared** radiation and **microwaves** are associated with **molecules**. **Radio waves** come from oscillations in **electric fields**.

Type	Approximate wavelength / m	Penetration	Uses	Effect on the Human Body
Radio waves	$10^{-1} — 10^6$	Pass through matter.	Radio transmissions.	No effect.
Microwaves	$10^{-3} — 10^{-1}$	Mostly pass through matter, but cause some heating.	Radar. Microwave cookery. TV transmissions.	Absorbed by water — danger of cooking human body.
Infrared (IR)	$7 \times 10^{-7} — 10^{-3}$	Mostly absorbed by matter, causing it to heat up.	Heat detectors. Night vision cameras. TV remote controls. Optical fibres.	Heating. Excess heat can harm the body's systems.
Visible light	$4 \times 10^{-7} — 7 \times 10^{-7}$	Absorbed by matter, causing some heating.	Human sight. Optical fibres.	Too bright a light can damage eyes.
Ultraviolet (UV)	$10^{-9} — 4 \times 10^{-7}$	Absorbed by matter. Slight ionisation.	Sunbeds. Security markings that show up in UV light.	Can cause skin cancer and eye damage.
X-rays	$10^{-13} — 10^{-9}$	Mostly pass through matter, but cause ionisation as they pass.	Medical/dental diagnosis. Airport security scanners. To kill cancer cells.	Can cause cancer due to cell damage.
Gamma rays	$10^{-16} — 10^{-10}$	Mostly pass through matter, but cause ionisation as they pass.	Irradiation of food. Sterilisation of medical instruments. To kill cancer cells.	Can cause cancer due to cell damage.

Electromagnetic Waves and Intensity

This page is for Edexcel only.

Intensity is a Measure of How Much Energy a Wave is Carrying

1) **Sound**, **light** and other waves all **carry energy**.

2) When you talk about the "**brightness**" of a light or "**loudness**" of a sound, what you really mean is **how much light** or **sound** energy is hitting your eyes or your ears **per second**.

3) The scientific measure of this is **intensity**.

> Intensity is the **rate of flow** of energy per **unit area** at **right angles** to the **direction of travel** of the wave. It's measured in **Wm^{-2}**.

4) Intensity is directly proportional to the **square of the amplitude** of the wave: $I \propto A^2$

Energy Spreading From a Point Source Forms a Sphere

1) The **power P** of a source is the **energy** it emits **per second**.

2) If you have waves **radiating** evenly from a **point source**, then at a distance r from the source, P is spread out over a **sphere** of surface area $4\pi r^2$.

3) So intensity I at a distance r from the source is given by:

$$I = \frac{P}{4\pi r^2}$$

4) Notice that I is **directly proportional** to $1/r^2$. If you move **twice as far** from a light source the light is **one quarter** as intense. **Three times** further out the light is **one ninth** as intense.

5) **Inverse square laws** crop up all over the place in physics.

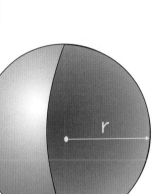

Practice Questions

Q1   What are the main practical uses of infrared radiation?

Q2   Why are microwaves dangerous?

Q3   How does the energy of an EM wave vary with frequency?

Q4   What are the SI units of intensity?

Q5   If P is the total energy emitted from a point source per second, what is the formula for the intensity at a distance r from the source?

Exam Questions

Q1   In a vacuum, do X-rays travel faster, slower or at the same speed as visible light? Explain your answer.   [2 marks]

Q2   (a) How can X-rays be detected?   [1 mark]

(b) Describe briefly the physics behind a practical use of X-rays.   [2 marks]

(c) What is the difference between gamma rays and X-rays?   [2 marks]

Q3   Give an example of a type of electromagnetic wave causing a hazard to health.   [2 marks]

Hey — without EM waves you wouldn't be able to see to revise...

There's not much physics on this page apart from the inverse square law bit. It's mainly just loads of facts to learn on the electromagnetic spectrum. You'll probably recognise most of it from GCSE, but you need to know it all well enough to be able to answer an exam question on it. So come on then, get to it...

Superposition and Coherence

These pages are for Edexcel and AQA A only.

When two waves get together, it can be either really impressive or really disappointing.

Superposition Happens When Two or More Waves Pass Through Each Other

1) At the **instant** when the waves **cross**, the **displacements** due to each wave **combine**. Then **each wave** goes on its merry way. You can **see** this happening if **two pulses** are sent **simultaneously** from each end of a rope or slinky.

2) The **principle of superposition** says that when two or more **waves cross**, the **resultant** displacement equals the **vector sum** of the **individual** displacements.

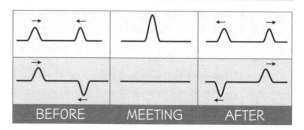

BEFORE   MEETING   AFTER

"Superposition" means "one thing on top of another thing." You can use the same idea in reverse — a complex wave can be separated out mathematically into several simple sine waves of various sizes.

Interference can be Constructive or Destructive

1) A **crest** plus a **crest** gives a **big crest**. A **trough** plus a **trough** gives a **big trough**. These are both examples of **constructive interference**.

2) A **crest** plus a **trough** of equal size gives... **nothing**. The two displacements **cancel each other out** completely. This is called **destructive interference**.

3) If the **crest** and the **trough** aren't the **same size**, then the destructive interference **isn't total**. For the interference to be **noticeable**, the two **amplitudes** should be **nearly equal**.

Graphically, you can superimpose waves by adding together the individual displacements at each point along the x-axis, and then plotting them.

You Get Patterns of Loud and Quiet from Interference of Sound Waves

Set up **two speakers** attached to the **same sound source** at **either end** of a long bench. If you walk from one end of the bench to the other, you'll hear the sound getting **louder** then **quieter** then **louder** again.

What you're hearing is **bands** of **constructive** and **destructive** interference, depending on whether the waves are **in phase** or **out of phase**.

In Phase Means In Step — Two Points In Phase Interfere Constructively

1) Two points on a wave are **in phase** if they are both at the **same point** in the **wave cycle**. Points in phase have the **same displacement** and **velocity**. In the graph below, points **A** and **B** are **in phase**; points **A** and **C** are **out of phase**.

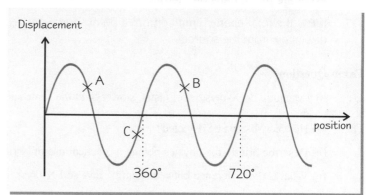

2) It's mathematically **handy** to show one **complete cycle** of a wave as an **angle of 360°**. **Two points** with a **phase difference** of **zero** or a **multiple of 360°** are **in phase**. **Points** with a **phase difference** of odd-number multiples of 180° are **exactly out of phase**.

3) You can also talk about two **different waves** being **in phase**. **In practice** this happens because **both** waves **originally** came from the **same oscillator**. In **other** situations there will nearly always be a **phase difference** between the two waves.

Superposition and Coherence

To Get Interference Patterns the Two Sources Must Be Coherent

Interference **still happens** when you're observing waves of **different wavelength** and frequency — but it happens in a **jumble**. In order to get clear **interference patterns**, the two or more sources must be **coherent**.

Two sources are **coherent** if they have the **same wavelength** and **frequency** and a **fixed phase difference** between them.

In exam questions at A2, the 'fixed phase difference' is almost certainly going to be zero. The two sources will be in phase.

Constructive or Destructive Interference Depends on the Path Difference

1) Whether you get **constructive** or **destructive** interference at a **point** depends on how **much further one wave** has travelled than the **other wave** to get to that point.

2) The **amount** by which the path travelled by one wave is **longer** than the path travelled by the other wave is called the **path difference**.

3) At **any point an equal distance** from both sources you will get **constructive interference**. You also get constructive interference at any point where the **path difference** is a **whole number of wavelengths**. At these points the two waves are **in phase** and **reinforce** each other. But at points where the path difference is **half a wavelength**, **one and a half** wavelengths, **two and a half** wavelengths etc., the waves arrive **out of phase** and you get **destructive interference**.

Loud	Path diff = λ
Quiet	Path difference = $\frac{\lambda}{2}$
Loud	No path difference
Quiet	Path difference = $\frac{\lambda}{2}$
Loud	Path diff = λ

$\phi = \frac{2\pi x}{\lambda}$

$\phi = \Delta$phase

$x = \Delta$path

Constructive interference occurs when: path difference = $n\lambda$ (where **n** is an integer)

Destructive interference occurs when: path difference = $\frac{(2n+1)\lambda}{2} = (n + \frac{1}{2})\lambda$

Practice Questions

Q1 Why does the principle of superposition deal with the **vector** sum of two displacements?

Q2 What happens when a crest meets a slightly smaller trough?

Q3 If two points on a wave have a phase difference of 1440°, are they in phase?

Exam Questions

Q1 (a) Two sources are coherent. What can you say about their frequencies, wavelengths and phase difference? [2 marks]

(b) Suggest why you might have difficulty in observing interference patterns in an area affected by two waves from two sources even though the two sources are coherent. [1 mark]

Q2 Two points on an undamped wave are exactly out of phase.

(a) What is the phase difference between them, expressed in degrees? [1 mark]

(b) Compare the displacements and velocities of the two points. [2 marks]

Learn this and you'll be in a super position to pass your exam... ...I'll get my coat.

There are a few really crucial concepts here: a) interference can be constructive or destructive, b) constructive interference happens when the path difference is a whole number of wavelengths, c) you need to use coherent sources.

Standing Waves

These pages are for Edexcel and AQA A only.

Standing waves are waves that... er... stand still... well, not still exactly... I mean, well... they don't go anywhere... um...

You get Standing Waves When a **Progressive Wave** is **Reflected** at a **Boundary**

A standing wave is the **superposition** of **two progressive waves** with the **same wavelength**, moving in **opposite directions**.

1) Unlike progressive waves, **no energy** is transmitted by a standing wave.

2) You can demonstrate standing waves by setting up a **driving oscillator** at one end of a **stretched string** with the other end fixed. The wave generated by the oscillator is **reflected** back and forth.

3) For most frequencies the resultant **pattern** is a **jumble**. However, if the oscillator happens to produce an **exact number of waves** in the time it takes for a wave to get to the **end** and **back again**, then the **original** and **reflected** waves **reinforce** each other.

4) At these **"resonant frequencies"** you get a **standing wave** where the **pattern doesn't move** — it just sits there, bobbing up and down. Happy, at peace with the world...

A sitting wave.

Standing Waves in **Strings** Form **Oscillating "Loops"** Separated by **Nodes**

1) Each particle vibrates at **right angles** to the string. **Nodes** are where the **amplitude** of the vibration is **zero**. **Antinodes** are points of **maximum amplitude**.

2) At resonant frequencies, an **exact number** of **half wavelengths** fits onto the string.

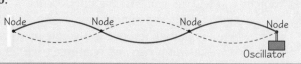

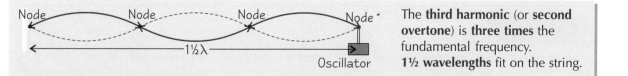

The standing wave above is vibrating at the **lowest possible** resonant frequency (the **fundamental frequency**). It has **one** "loop" with a **node at each end**.

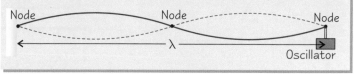

This is the **second harmonic** (or **first overtone**). It is **twice** the fundamental frequency. There are two **"loops"** with a **node** in the **middle** and **one at each end**.

The **third harmonic** (or **second overtone**) is **three times** the fundamental frequency. **1½ wavelengths** fit on the string.

The **Notes** Played by **Stringed** and **Wind Instruments** are Standing Waves

Transverse standing waves form on the strings of **stringed instruments** like **violins** and **guitars**. Your finger or the bow sets the **string vibrating** at the point of contact. Waves are sent out in **both directions** and **reflected** back at both ends.

Longitudinal Standing Waves Form in a **Wind Instrument** or Other **Air Column**

1) If a source of sound is placed at the open end of a flute, piccolo, oboe or other column of air, there will be some **frequencies** for which **resonance** occurs and a standing wave is set up.

2) If the instrument has a **closed end**, a **node** will form there. You get the lowest resonant frequency when the length, *l*, of the pipe is a **quarter wavelength**.

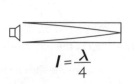

$l = \dfrac{\lambda}{4}$

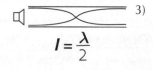

$l = \dfrac{\lambda}{2}$

3) **Antinodes** form at the **open ends** of pipes. If both ends are open, you get the lowest resonant frequency when the length, *l*, of the pipe is a **half wavelength**.

Remember, the sound waves in wind instruments are <u>longitudinal</u> — they <u>don't</u> actually look like these diagrams.

Standing Waves

Fundamental Frequency of a String Depends on Length, Weight and Tension

1) The **longer** the string, the **lower** the note — because the **half wavelength** at the natural frequency is longer.

2) The **heavier** (i.e. the more mass per length) the string, the **lower** the note — because waves travel more **slowly** down the string. For a given **length** a **lower** velocity, *v*, makes a **lower** frequency, *f*.

3) The **looser** the string, the **lower** the note — again because waves travel more **slowly** down a **loose** string.

You can Demonstrate Standing Waves with Microwaves and Sounds

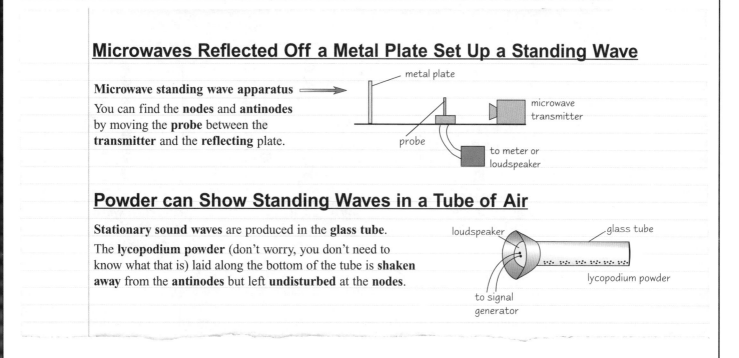

Microwaves Reflected Off a Metal Plate Set Up a Standing Wave

Microwave standing wave apparatus
You can find the **nodes** and **antinodes** by moving the **probe** between the **transmitter** and the **reflecting** plate.

Powder can Show Standing Waves in a Tube of Air

Stationary sound waves are produced in the **glass tube**.

The **lycopodium powder** (don't worry, you don't need to know what that is) laid along the bottom of the tube is **shaken away** from the **antinodes** but left **undisturbed** at the **nodes**.

Practice Questions

Q1 How do standing waves form?

Q2 At four times the fundamental frequency, how many half wavelengths fit on a violin string?

Q3 What factors affect the fundamental frequency of a guitar string?

Q4 Describe an experiment to investigate standing waves in a column of air.

Exam Questions

Q1 (a) A standing wave of three times the fundamental frequency is formed on a stretched string of length 1.2 m. Sketch a diagram showing the form of the wave. [2 marks]

(b) What is the wavelength of the standing wave? [1 mark]

(c) Explain how the amplitude varies along the string. How is that different from the amplitude of a progressive wave? [2 marks]

(d) At a given moment, how does the displacement of a particle at one antinode compare to the displacement of a particle at the next antinode? [2 marks]

CGP — putting the FUN back in FUNdamental frequency...

Resonance was a big problem for the Millennium Bridge in London. The resonant frequency of the bridge was round about normal walking pace, so as soon as people started using it they set up a huge standing wave. An oversight, I feel...

Diffraction

These pages are for Edexcel and AQA A only.

Ripple tanks, ripple tanks — yeah.

Waves Go **Round Corners** and **Spread out** of **Gaps**

The way that **waves spread out** as they come through a **narrow gap**, or go round obstacles, is called **diffraction**. **All** waves diffract, but it's not always easy to observe.

Diffraction of Water Waves using a Ripple Tank

You can make diffraction patterns in ripple tanks.
The **amount** of diffraction depends on the **wavelength** of the wave compared with the **size of the gap**.

These are <u>wavefronts</u>.
Edexcel: by definition,
all points on a
wavefront are <u>in phase</u>
(see p. 24).

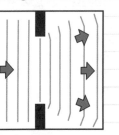
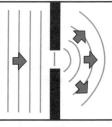

When the **gap** is **a lot bigger** than the wavelength, diffraction is **unnoticeable**.

You get **noticeable diffraction** through a gap **several wavelengths wide**.

You get the **most** diffraction when the gap is **the same size** as the wavelength.

If the gap is **smaller** than the wavelength, the waves are just **reflected back**.

When **sound** passes through a **doorway**, the **size of gap** and the **wavelength** are about **equal**, so the **maximum diffraction** occurs. That's why you have no trouble **hearing** someone through an **open door** to the next room, even if the other person is out of your **line of sight**. The reason that you can't **see** him or her is that when **light** passes through the doorway, it is passing through a **gap** around a **hundred million times bigger** than its wavelength — the amount of diffraction is **tiny**.

You can Demonstrate **Diffraction** in **Light** Using a **Laser**

1) Diffraction in **light** can be demonstrated by shining **laser light** through a very **narrow slit** onto a screen. You can alter the amount of diffraction by changing the width of the slit.

2) You can do a similar experiment using a **white light** source instead of the laser (which is monochromatic) and a set of **colour filters**. The size of the slit can be kept constant while the **wavelength** is varied by putting different **colour filters** over the slit.

Warning. Use of coloured filters may result in excessive fun.

Diffraction is Sometimes **Useful** and Sometimes a Pain...

1) For a **satellite transmission** you want a **narrow**, **targeted** beam, so satellite communications companies aim to **minimise** diffraction.

2) For a **loudspeaker** you want the sound to be heard as widely as possible, so you aim to **maximise** diffraction.

3) With a **microwave oven** you want to **stop** the **microwaves** diffracting out and frying your kidneys **and** you want to **let light through** so you can **see** your food. A **metal mesh** on the **door** has **gaps too small** for microwaves to diffract through, but **light** slips through because of its **much smaller wavelength**.

Diffraction

With **Light Waves** you get a **Pattern** of **Light** and **Dark Fringes**

1) If the wavelength of a light wave is about the same size as the aperture, you get a diffraction pattern of light and dark fringes.

2) The pattern has a bright central fringe with alternating dark and bright fringes on either side of it.

3) The first set of dark fringes are produced at an angle θ to the direction of the incident light, where:

$$\sin \theta = \frac{\lambda}{d}$$

You have to be able to recall and use this formula, but you don't have to know why it works.

Diffraction Puts a **Limit** on How Much **Detail** You Can See

1) **Blurring** due to **diffraction** means that if you're looking at **two points** it can be hard to **resolve** one from the other. The **limit of resolution** is when you can't tell any more whether you're looking at **two sources** or **one**.

2) When the **wavelength** is **much smaller** than the **gap** (e.g. the lens of a telescope), there's hardly any **diffraction** and resolution is **good**. As the **wavelength** gets **closer** to the size of the gap, diffraction **increases** and resolution gets **worse**.

3) This is a **big problem** in **astronomy**. Even the **best** telescopes can't **resolve two stars** very close on the sky.

4) One solution is to make the "**gap**" **very large indeed** by building an **array** of telescopes that together make an aperture **several kilometres wide**. Talk about extreme measures...

5) Diffraction also causes problems in **resolving details** in an **ultrasound** image. To minimise diffraction, it's best to have the **wavelength** as **small** as possible.

Practice Questions

Q1 What is diffraction?

Q2 Sketch what happens when plane waves meet an obstacle about as wide as one wavelength.

Q3 For a long time some scientists argued that light couldn't be a wave because it did not seem to diffract. Why did they get this impression?

Q4 Do all waves diffract?

Q5 If the wavelength of light causing a diffraction pattern gets bigger, does the pattern get wider or narrower?

Exam Questions

Q1 A mountain lies directly between you and a radio transmitter. Explain using diagrams why you can pick up long-wave radio broadcasts from the transmitter but not short-wave. [4 marks]

Q2 A beam of coherent light of wavelength 6×10^{-7} m falls on a narrow slit of width 0.1 mm producing a diffraction pattern on a screen some metres away.

(a) State qualitatively what would happen to the width of the pattern if the slit were halved in width. [1 mark]

(b) Calculate the angle θ the first dark fringe makes with the incident beam. [2 marks]

Even hiding behind a mountain, you can't get away from long-wave radio...

*Diffraction crops up again in particle physics, astronomy and quantum physics, so you **really** need to understand it. Make sure you learn that equation at the top of the page — you won't be given it in the exam.*

Two-Source Interference

These pages are for Edexcel and AQA A only.

Yeah, I know, fringe spacing doesn't really sound like a Physics topic — just trust me on this one, OK.

Demonstrating **Two-Source** Interference in **Water** and **Sound** is Easy

1) It's **easy** to demonstrate **two-source interference** for either **sound** or **water** because they've got **wavelengths** of a handy **size** that you can **measure**.

2) You need **coherent** sources, which means the **wavelength** and **frequency** have to be the **same**. The trick is to use the **same oscillator** to drive **both sources**. For **water**, one **vibrator** drives two **dippers**. For sound, **one oscillator** is connected to **two loudspeakers**. (See diagram on page 25.)

Demonstrating **Two-Source** Interference for **Light** is Harder

Young's Double-Slit Experiment

1) You **can't** arrange **two separate coherent light sources** because **light** from **each source** is emitted in **random bursts**. Instead a **single** laser is shone through **two slits**.

2) Light spreading out by **diffraction** from the slits is equivalent to **two coherent point sources**. Laser light is used so that there's only **one wavelength** present.

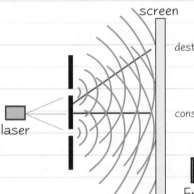

Fringes on screen

3) You get a pattern of light and dark **fringes**, depending whether constructive or destructive **interference** is taking place. Thomas Young — the first person to do this experiment — came up with an **equation** to **work out** the **wavelength** of the **light** from this experiment (see next page).

4) This is a modern version of **Young's Experiment**. The original experiment was important because it showed that **light has the properties of a wave**. **Interference** and **diffraction** are both **wave** properties and **can't** be explained any other way. (See p. 60.)

5) However, other experiments show that light **also** has the properties of a particle. This weird **wave-particle duality** comes up later in the book.

You can do a similar experiment with microwaves instead of visible light. If it tickles your hamster...

Two-Source Interference

Work Out the Wavelength with Young's Double-Slit Formula

1) The fringe spacing (**X**), wavelength (**λ**), spacing between slits (**d**) and the distance from slits to screen (**D**) are all related by **Young's double-slit formula** which you need to know but not prove:

$$\text{Fringe spacing, } X = \frac{D\lambda}{d}$$

"Fringe spacing" means the distance from the centre of one minimum to the centre of the next minimum or from the centre of one maximum to the centre of the next maximum.

2) Since the wavelength is so small you can see from the formula that a high ratio of **D / d** is needed to make the fringe spacing **big enough to see**.

3) Rearranging you can use **λ = Xd / D** to **calculate the wavelength** of light.

4) The fringes are **so tiny** that it's very hard to get an **accurate value of X**. It's easier to measure across **several** fringes then **divide** by the number of **fringe widths** between them.

Always check your fringe spacing.

You Can Also Get Interference Patterns Using a Partial Reflector

1) Light sent in at an **angle** is **partially transmitted** and **partially reflected** at the **half-silvered surface** of the **partial** reflector.

2) The reflected beam **bounces back** from the **left hand reflector** and the same thing happens **again** repeatedly.

3) The end result is **multiple parallel beams**, with each beam **out of phase** with the next by a **fixed amount**.

4) The beams can be focused with a lens to produce an **interference pattern**.

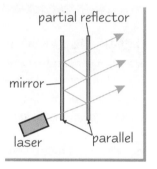

partial reflector

mirror

laser   parallel

Practice Questions

Q1 In Young's experiment, why do you get a bright fringe at a point equidistant from both slits?

Q2 What does Young's experiment show about the nature of light?

Q3 Write down Young's double-slit formula.

Exam Questions

Q1 (a) The diagram on the right shows waves from two coherent light sources S_1 and S_2. Sketch the interference pattern marking on constructive and destructive interference.
[2 marks]

S_1•)
S_2•)

(b) In practice if interference is to be observed S_1 and S_2 must be slits in a screen behind which there is a source of laser light. Why?
[2 marks]

Q2 In an experiment to study sound interference, two loudspeakers are connected to an oscillator emitting sound at 1320 Hz and set up as shown in the diagram below. They are 1.5 m apart and 7 m away from the line AC. A listener moving from A to C hears minimum sound at A and C and maximum sound at B.

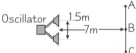

Oscillator   1.5m
            7m   B
                      C
                     A

(a) Calculate the wavelength of the sound waves if the speed of sound in air is taken to be 330 ms^{-1}. [1 mark]

(b) Calculate the separation of points A and C. [2 marks]

Carry on Physics — this page is far too saucy...

(Ooh er...)

Be careful when you're calculating the fringe width by averaging over several fringes. Don't just divide by the number of bright lines. Ten bright lines will only have nine fringe-widths between them, not ten. It's an easy mistake to make, but you have been warned... mwa ha ha ha (felt necessary, sorry).

Diffraction Gratings

These pages are for AQA A only.

Ay... starting to get into some pretty funky stuff now. I like light experiments.

Interference Patterns Get **Sharper** When You Diffract Through **More Slits**

1) You can repeat **Young's double-slit** experiment (see p. 30) with **more than two equally spaced** slits. You get basically the **same shaped** pattern as for two slits — but the **bright bands** are **brighter** and **narrower** and the **dark areas** between are **darker**.

2) When **monochromatic light** (one wavelength) is passed through a **grating** with **hundreds** of slits per millimetre, the interference pattern is **really sharp** because there are so **many beams reinforcing** the **pattern**.

3) Sharper fringes make for more **accurate** measurements.

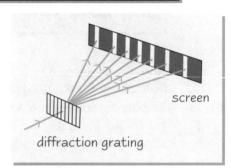

diffraction grating

screen

Monochromatic Light on a **Diffraction Grating** gives **Sharp Lines**

1) For **monochromatic** light, all the **maxima** are sharp lines. (It's different for white light — see p.33.)

2) There's a line of **maximum brightness** at the centre called the **zero order** line.

3) The lines just **either side** of the central one are called **first order lines**. The **next pair out** are called **second order** lines and so on.

4) For a grating with slits a distance *d* apart, the angle between the **incident beam** and **the nth order maximum** is given by:

$$d \sin \theta = n\lambda$$

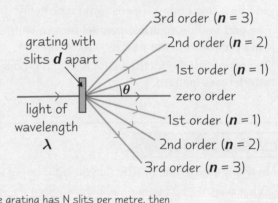

grating with slits *d* apart

light of wavelength λ

3rd order (*n* = 3)
2nd order (*n* = 2)
1st order (*n* = 1)
zero order
1st order (*n* = 1)
2nd order (*n* = 2)
3rd order (*n* = 3)

5) So by observing *d*, *θ* and *n* you can **calculate the wavelength** of the light.

If the grating has N slits per metre, then the slit spacing, d, is just 1/N metres.

DERIVING THE EQUATION:

1) At **each slit**, the incoming waves are **diffracted**. These diffracted waves then **interfere** with each other to produce an **interference pattern**.

2) Consider the **first order maximum**. This happens at the **angle** when the waves from one slit line up with waves from the **next slit** that are **exactly one wavelength** behind.

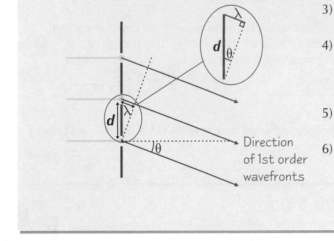

Direction of 1st order wavefronts

3) Call the **angle** between the **first order maximum** and the **incoming light** θ.

4) Now, look at the **triangle** highlighted in the diagram. The angle is θ (using basic geometry), *d* is the slit spacing and the **path difference** is λ.

5) So, for the first maximum, using trig:
$$d \sin \theta = \lambda$$

6) The other maxima occur when the path difference is 2λ, 3λ, 4λ, etc. So to make the equation **general**, just replace λ with *n*λ, where *n* is an integer — the **order** of the maximum.

Diffraction Gratings

You can Draw General Conclusions from d sin θ = nλ

1) If λ is **bigger**, sin θ is **bigger**, and so θ is **bigger**. This means that the larger the **wavelength**, the more the pattern will **spread out**.

2) If d is **bigger**, sin θ is **smaller**. This means that the **coarser** the **grating**, the **less** the pattern will **spread out**.

3) Values of sin θ greater than **1** are **impossible**. Light **can't** be scattered by an angle of more than **90°**. If for a certain n you get a result of **more than 1** for sin θ you know that order **doesn't exist**.

Shining White Light Through a Diffraction Grating Produces Spectra

1) **White light** is really a **mixture** of **colours**. If you **diffract** white light through a **grating** then the patterns due to **different wavelengths** within the white light are **spread out** by **different** amounts.

2) Each **order** in the pattern becomes a **spectrum**, with **red** on the **outside** and **violet** on the **inside**. The **zero order maximum** stays **white** because there's **no path difference** between waves arriving there.

second order   first order   zero order   first order   second order
(white)

Astronomers and chemists often need to study spectra. They use diffraction gratings rather than prisms because they're more accurate.

Practice Questions

Q1 How is the diffraction grating pattern for white light different from the pattern for laser light?

Q2 What difference does it make to the pattern if you use a finer grating?

Q3 What equation is used to find the angle between the nth order maximum and the incident beam for a diffraction grating?

Q4 Derive the equation you quoted in Q3.

Exam Questions

Q1 Yellow laser light of wavelength 600 nm (6×10^{-7} m) is transmitted through a diffraction grating of 4×10^5 lines per metre.

(a) At what angle to the normal are the first and second order bright lines seen? [4 marks]

(b) Is there a fifth order line? [1 mark]

Q2 Visible, monochromatic light is transmitted through a diffraction grating of 3.7×10^5 lines per metre. The first order maximum is at an angle of 14.2° to the incident beam.

Find the wavelength of the incident light. [2 marks]

Ooooooooooooooo — pretty patterns...

*Derivation — ouch. At least it's not a bad one though. As long as you learn the diagram, its just geometry and a bit of trig from there. Make sure you **learn** the equation — that way, you know what you're aiming for. As for the rest of the page, remember that the more slits you have, the sharper the image — and white light makes a pretty spectrum.*

Gravitational Fields

These pages are for Edexcel, AQA A, AQA B, OCR A and OCR B.

*Gravity's all about masses **attracting** each other. If the Earth didn't have a **gravitational field,** apples wouldn't fall to the ground and you'd probably be floating off into space instead of sitting here reading this page...*

Masses in a **Gravitational Field** experience a **Force Of Attraction**

Every object with mass has a **gravitational field**. This is the region where it can **exert** an **attractive force** on other masses, without touching them.

1)   Any object with mass will **experience a force** if you put it in the **gravitational field** of another object.

2)   All objects with mass **attract** each other.

3)   Only objects with a **large** mass, such as stars and planets, have a significant effect. E.g. the gravitational fields of the **Moon** and the **Sun** are noticeable here on Earth — they're the main cause of our **tides**.

You can draw **Lines of Force** to show the **Field** around an object

Lines of force (or "**field lines**") are **arrows** showing the **direction of the force** that masses would feel in the gravitational field.

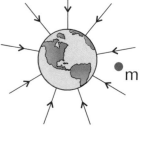

1)   If you put a small mass, *m*, anywhere in the Earth's gravitational field, it will always be attracted **towards** the Earth.

2)   The Earth's gravitational field is **radial** — the lines of force meet at the centre of the Earth.

3)   If you move mass *m* further away from the Earth — where the **lines** of force are **further apart** — the **force** it experiences **decreases**.

4)   The small mass, *m*, has a gravitational field of its own. This doesn't have a noticeable effect on the Earth though, because the Earth is so much **more massive**.

The **Field Strength** is the **Force per Unit Mass**

Gravitational field strength, *g*, is the **force per unit mass**. Its value depends on **where you are** in the field. There's a really simple equation for working it out:

$$g = \frac{F}{m}$$

1)   *F* is the force experienced by a mass *m* when it's placed in the gravitational field. Divide *F* by *m* and you get the **force per unit mass**.

2)   *g* is a **vector** quantity, always pointing towards the centre of the mass whose field you're describing.

3)   The units of *g* are **newtons per kilogram** (Nkg^{-1}).

You can **Calculate Forces** using **Newton's Law of Gravitation**

The **force** experienced by an object in a gravitational field is always **attractive**. It's a **vector** which depends on the **masses** involved and the **distances** between them. It's easy to work this out for **point masses** — objects which behave as if all their mass is concentrated at the centre. You just put the numbers into this equation...

NEWTON'S LAW OF GRAVITATION:

$$F = (-)\frac{Gm_1m_2}{r^2}$$

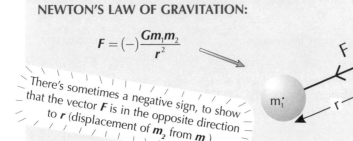

There's sometimes a negative sign, to show that the vector *F* is in the opposite direction to *r* (displacement of *m₂* from *m₁*).

The diagram shows the force acting on m_2 due to m_1. (The force on m_1 due to m_2 is equal but in the other direction.)

m_1 and m_2 are point masses.

G is the **gravitational constant** — $6.67 \times 10^{-11}\ Nm^2kg^{-2}$.

r is the distance (in metres) between the centres of the two masses.

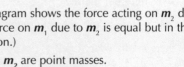

If you're doing OCR B, the m_1 and m_2 would be replaced by *M* and *m*

The law of gravitation is an **inverse square law** $\left(F \propto \frac{1}{r^2}\right)$ so:

1)   if the distance **r** between the masses **increases** then the force **F** will **decrease**.

2)   if the **distance doubles** then the **force** will be one **quarter** the strength of the original force.

Gravitational Fields

In a *Radial Field*, *g* is *Inversely Proportional* to *r²*

Point masses have **radial** gravitational fields. The value of **g** depends on the distance **r** from the centre of the point mass **M**...

$$g = \frac{GM}{r^2}$$

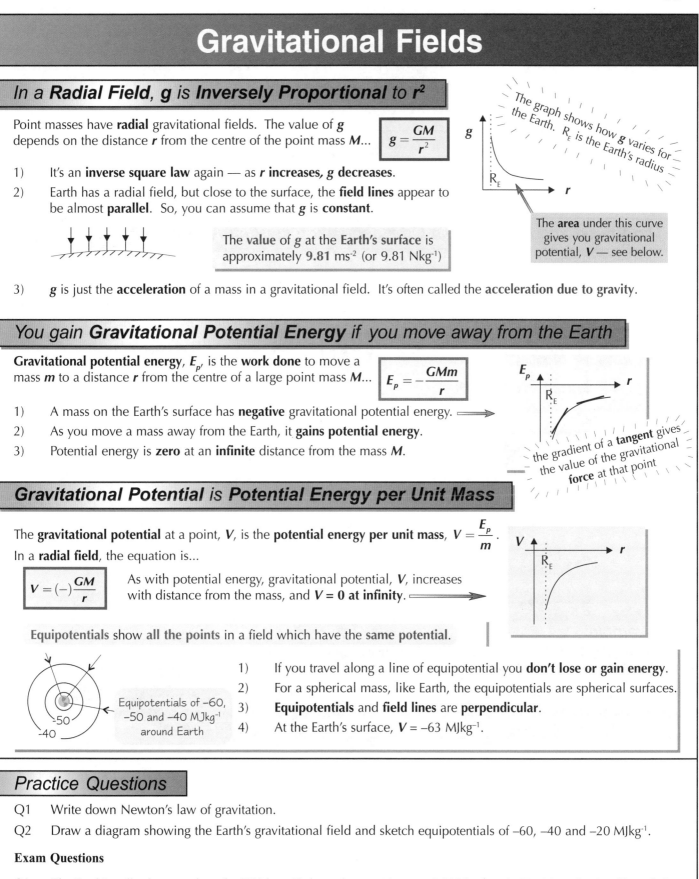

*The graph shows how **g** varies for the Earth. R_E is the Earth's radius*

1) It's an **inverse square law** again — as **r** increases, **g** decreases.
2) Earth has a radial field, but close to the surface, the **field lines** appear to be almost **parallel**. So, you can assume that **g** is **constant**.

The **value of g** at the **Earth's surface** is approximately **9.81** ms^{-2} (or 9.81 Nkg^{-1})

The **area** under this curve gives you gravitational potential, **V** — see below.

3) **g** is just the **acceleration** of a mass in a gravitational field. It's often called the **acceleration due to gravity**.

You gain *Gravitational Potential Energy* if you move away from the Earth

Gravitational potential energy, E_p, is the **work done** to move a mass **m** to a distance **r** from the centre of a large point mass **M**...

$$E_p = -\frac{GMm}{r}$$

1) A mass on the Earth's surface has **negative** gravitational potential energy.
2) As you move a mass away from the Earth, it **gains potential energy**.
3) Potential energy is **zero** at an **infinite** distance from the mass **M**.

*the gradient of a **tangent** gives the value of the gravitational **force** at that point*

Gravitational Potential is *Potential Energy per Unit Mass*

The **gravitational potential** at a point, **V**, is the **potential energy per unit mass**, $V = \frac{E_p}{m}$.
In a **radial field**, the equation is...

$$V = (-)\frac{GM}{r}$$

As with potential energy, gravitational potential, **V**, increases with distance from the mass, and **V = 0 at infinity**.

Equipotentials show **all the points** in a field which have the **same potential**.

Equipotentials of −60, −50 and −40 $MJkg^{-1}$ around Earth

1) If you travel along a line of equipotential you **don't lose or gain energy**.
2) For a spherical mass, like Earth, the equipotentials are spherical surfaces.
3) **Equipotentials** and **field lines** are **perpendicular**.
4) At the Earth's surface, **V** = −63 $MJkg^{-1}$.

Practice Questions

Q1 Write down Newton's law of gravitation.

Q2 Draw a diagram showing the Earth's gravitational field and sketch equipotentials of −60, −40 and −20 $MJkg^{-1}$.

Exam Questions

Q1 The Earth's radius is approximately 6400 km. Estimate its mass (use g = 9.81 Nkg^{-1} at the Earth's surface). [2 marks]

Q2 The moon has mass 7.43×10^{22} kg and radius 1740 km.

(a) Calculate the value of **g** at the Moon's surface. [1 mark]

(b) Calculate the gravitational potential energy of a 25 kg mass at a height of 10 km above the Moon's surface. [2 marks]

If you're really stuck, put 'Inverse Square Law'...

Clever chap, Newton, but famously tetchy. He got into fights with other physicists, mainly over planetary motion and calculus... the usual playground squabbles. Then he spent the rest of his life trying to turn scrap metal into gold. Weird.

Motion of Masses in Gravitational Fields

This page is for Edexcel, AQA A, AQA B, OCR A Option 01 and OCR B.

*Planets just go round and round in circles. Well, **ellipses** really, but I won't tell if you don't...*

Planets *are* Satellites *which* Orbit *the* Sun

1) A **satellite** is just any **smaller mass** which **orbits** a **much larger mass** — the **Moon** is the Earth's satellite.

2) In our solar system, the planets have **nearly circular orbits...** so you can use the **equations of circular motion**.

The *Speed* and *Period* of an Orbit depend on its *Radius* and the Mass of the *Sun*

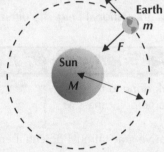

1) Earth feels a force due to the gravitational 'pull' of the **Sun**. This force is given by Newton's law of gravitation...

$$F = \frac{GMm}{r^2}$$ (see p. 34)

2) The Earth has velocity **v**. Its linear speed is constant but its **direction** is not — so it's accelerating. The **centripetal force** causing this acceleration is:

$$F = \frac{mv^2}{r}$$

2) The **centripetal force** on the Earth must be a result of the **gravitational force** due to the Sun, and so these forces must be **equal**...

$$\frac{mv^2}{r} = \frac{GMm}{r^2}$$ and rearranging... $$v = \sqrt{\frac{GM}{r}}$$

The **time** taken **for one orbit** is called the **period**, **T**. For circular motion, $T = \frac{2\pi r}{v}$.

Substitute for **v** and rearrange... $$T = \sqrt{\frac{4\pi^2 r^3}{GM}}$$

Example

The Moon takes 27.3 days to orbit the Earth. Calculate its distance from the Earth.
Take the mass of the Earth to be 5.975×10^{24} kg.

You're trying to find the radius of the orbit, **r**. Use the formula for period, **T**:

$$T = \sqrt{\frac{4\pi^2 r^3}{GM}}$$

You've been given the values of **T** (27.3 days) and **M**, and you'll be able to look up the value of **G** on the exam data sheet — $G = 6.67 \times 10^{-11}$ Nm²kg⁻².

and rearrange it for **r³**: $$r^3 = \frac{T^2 GM}{4\pi^2}$$

T = 27.3 days = 2.36×10^6 s
G = 6.67×10^{-11} Nm²kg⁻²
M = 5.975×10^{24} kg

and put the numbers in (convert to SI units first): $$r^3 = \frac{(2.36 \times 10^6)^2 \times (6.67 \times 10^{-11}) \times (5.975 \times 10^{24})}{4\pi^2} = 5.62 \times 10^{25}$$

$$r = 3.83 \times 10^8 \text{ m}$$
$$= \underline{\mathbf{3.83 \times 10^5 \text{ km}}} \text{ (this is the distance between the centre of the Earth and the centre of the Moon)}$$

Geosynchronous Satellites orbit the Earth once in **24 hours**

1) Geosynchronous satellites orbit directly over the **equator** and are **always above the same point** on Earth.

2) A geosynchronous satellite travels at the **same angular speed as the Earth** turns below it.

3) These satellites are really useful for sending TV and telephone signals — the satellite is **stationary** relative to a certain point on the **Earth**, so you don't have to alter the angle of your receiver (or transmitter) to keep up.

4) Their orbit takes exactly **one day**.

Motion of Masses in Gravitational Fields

This page is for OCR A Option 01 only.

Kepler's Laws are about the Motion of Planets in the Solar System

Kepler came up with these laws around 1600, about 80 years before Newton developed his Law of Gravitation.

1) Each planet moves in an **ellipse** around the Sun (a circle is just a special kind of ellipse).

2) Lines drawn from the Sun to each planet will sweep out **equal areas in equal times**.
If you assume that the orbits are **circular**, this means that the planets have **constant angular speed** — see p. 12.

3) The **period** of the orbit and the **mean distance** between the Sun and the planet are related by $T^2 \propto r^3$.

For circular motion, where r = radius of the orbit, $\dfrac{r^3}{T^2} = \text{constant} = \dfrac{GM}{4\pi^2}$ (see below).

Kepler's Third Law provides Experimental Evidence for Newton's Law of Gravitation

Kepler's **third law** — $T^2 \propto r^3$ — is based on the analysis of **real data** that was collected by a fellow scientist, **Brahe**.

It implies that, for a circular orbit with radius r, $\dfrac{r^3}{T^2}$ is a constant. And this supports Newton's law of gravitation...

Newton's law leads to this expression for the period of an orbit

$$T = \sqrt{\frac{4\pi^2 r^3}{GM}}$$

which can be rearranged to give

$$\frac{r^3}{T^2} = \frac{GM}{4\pi^2}$$

i.e.

$$\frac{r^3}{T^2} = \text{constant}$$

← which is exactly what Brahe and Kepler had concluded from their stargazing

Neptune's existence was predicted from the orbit of Uranus

1) Astronomers observed **disturbances** in the orbit of **Uranus** which couldn't be explained by the gravitational fields of the other known planets.

2) They concluded that there must be **another planet** out there causing the 'wobble'. They predicted its position and the size of its orbit using Newton's law of gravitation.

3) As a result of these predictions, Neptune was finally discovered in **1846**.

Practice Questions

Q1 Derive an expression for the radius of the orbit of a planet around the Sun, in terms of the period of its orbit.

Q2 The International Space Station orbits the Earth with velocity *v*. If another vehicle docks with it, increasing its mass, what difference, if any, does this make to the speed or radius of the orbit?

Q3 Would a geosynchronous satellite be useful for making observations for weather forecasts? Give reasons.

Q4 State Kepler's laws.

Exam Questions

(Use G = 6.67 × 10⁻¹¹Nm²kg⁻², mass of Earth = 5.98 × 10²⁴ kg, radius of Earth = 6400 km)

Q1 (a) A satellite orbits 50 km above the Earth's surface. Calculate the period of the satellite's orbit. [2 marks]

(b) What is the linear speed of the satellite? [1 mark]

Q2 At what height above the Earth's surface would a geosynchronous satellite orbit? [3 marks]

Here comes the Sun, little darling...

Galileo got into bother back in 1633 for supporting Copernicus and his wacky idea that the Earth moves round the Sun... The Catholic Church said that everything goes round the Earth, and after a few espressos with those nice chaps from the Inquisition, Galileo did too. In 1992, the Vatican admitted that maybe Galileo was right after all, and pardoned him.

Electric Fields

These pages are for Edexcel, AQA A, AQA B, OCR A and OCR B.

*Electric fields can be attractive or repulsive, so they're different from gravitational ones. It's all to do with **charge**.*

There is an **Electric Field** around a **Charged Object**

Any object with **charge** has an **electric field** around it — the region where it can attract or repel other charges.

1) Electric charge, **Q**, is measured in **coulombs** (C) and can be either positive or negative.
2) **Oppositely** charged particles **attract** each other. **Like** charges **repel**.
3) If a **charged object** is placed in an electric field, then it will experience a **force**.

You can **Calculate Forces** using **Coulomb's Law**

You'll need **Coulomb's law** to work out **F** — the force of attraction or repulsion between two point charges...

COULOMB'S LAW:

$$F = \frac{kQ_1Q_2}{r^2} \quad \text{where} \quad k = \frac{1}{4\pi\varepsilon}$$

ε = ("epsilon") permittivity of material between charges
Q_1 and Q_2 are the charges
r is the distance between Q_1 and Q_2

If the charges are **opposite** then the force is **attractive**. **F** will be **negative**.

If Q_1 and Q_2 are **like** charges then the force is **repulsive**, and **F** will be **positive**.

1) The force on Q_1 is always **equal** and **opposite** to the force on Q_2.
2) It's an **inverse square law**. Again. The further apart the charges are, the weaker the force between them.
3) The size of the force **F** also depends on the **permittivity**, ε, of the material between the two charges. For free space, the permittivity is $\varepsilon_0 = 8.85 \times 10^{-12}\,C^2N^{-1}m^{-2}$.

Electric Field Strength is Force per Unit Charge

Electric field strength, **E**, is defined as the **force per unit positive charge** — the force that a charge of +1 C would experience if it were placed in the electric field.

$$E = \frac{F}{q}$$

F is the force on a 'test' charge **q**

1) **E** is a **vector** pointing in the **direction** that a **positive charge** would **move**.
2) The units of **E** are **newtons per coulomb** (NC^{-1}).
3) Field strength depends on **where you are** in the field.
4) A **point charge** — which behaves as if all its charge is concentrated at the centre — has a **radial** field.

In a **Radial Field**, E is **Inversely Proportional** to r²

1) **E** is the force per unit charge that a small, positive 'test' charge, **q**, would feel at different points in the field. In a **radial field**, **E** depends on the distance **r** from the point charge **Q**...

$$E = \frac{kQ}{r^2} \quad \left(k = \frac{1}{4\pi\varepsilon}\right)$$

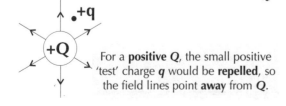

For a **positive Q**, the small positive 'test' charge **q** would be **repelled**, so the field lines point **away** from **Q**.

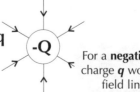

For a **negative Q**, the small positive charge **q** would be **attracted**, so the field lines point **towards Q**.

2) It's another **inverse square law** — $E \propto \dfrac{1}{r^2}$

3) Field strength **decreases** as you go **further away** from **Q** — on a diagram, the **field lines** get **further apart**.

Electric Fields

A Charge in an Electric Field has Electric Potential Energy

Electric potential energy, E_p, is the **work done** to move a small positively charged particle, q, from infinity to a distance, r, away from a point charge, Q...

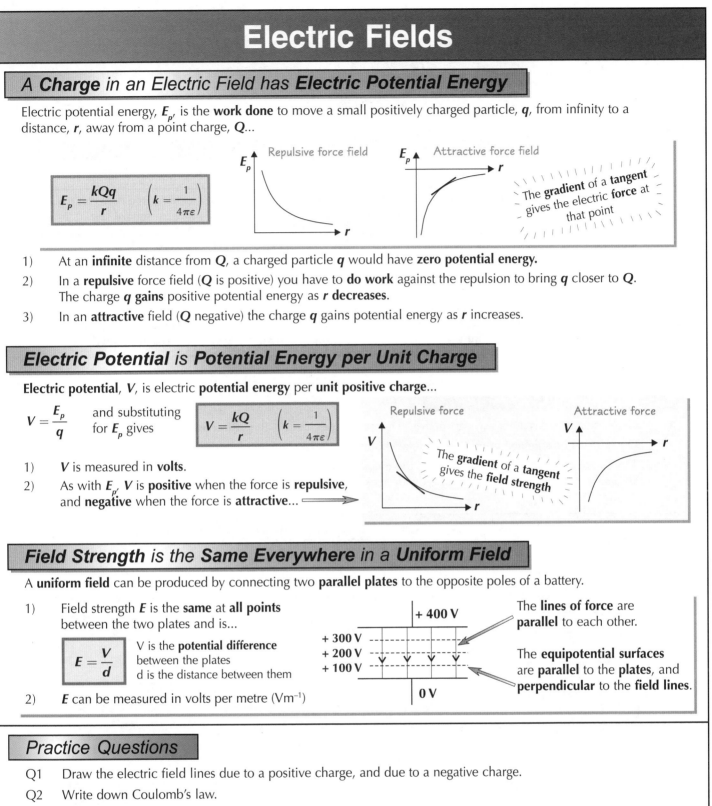

$$E_p = \frac{kQq}{r} \qquad \left(k = \frac{1}{4\pi\varepsilon}\right)$$

Repulsive force field

Attractive force field

The **gradient** of a **tangent** gives the electric **force** at that point

1) At an **infinite** distance from Q, a charged particle q would have **zero potential energy.**

2) In a **repulsive** force field (Q is positive) you have to **do work** against the repulsion to bring q closer to Q. The charge q **gains** positive potential energy as r **decreases.**

3) In an **attractive** field (Q negative) the charge q gains potential energy as r increases.

Electric Potential is Potential Energy per Unit Charge

Electric potential, V, is electric **potential energy** per **unit positive charge**...

$V = \frac{E_p}{q}$ and substituting for E_p gives

$$V = \frac{kQ}{r} \qquad \left(k = \frac{1}{4\pi\varepsilon}\right)$$

Repulsive force

Attractive force

The **gradient** of a **tangent** gives the **field strength**

1) V is measured in **volts**.

2) As with E_p, V is **positive** when the force is **repulsive**, and **negative** when the force is **attractive**...

Field Strength is the Same Everywhere in a Uniform Field

A **uniform field** can be produced by connecting two **parallel plates** to the opposite poles of a battery.

1) Field strength E is the **same** at **all points** between the two plates and is...

$$E = \frac{V}{d}$$

V is the **potential difference** between the plates
d is the distance between them

2) E can be measured in volts per metre (Vm^{-1})

+ 400 V

+ 300 V
+ 200 V
+ 100 V

0 V

The **lines of force** are **parallel** to each other.

The **equipotential surfaces** are **parallel** to the **plates**, and **perpendicular** to the **field lines**.

Practice Questions

Q1 Draw the electric field lines due to a positive charge, and due to a negative charge.

Q2 Write down Coulomb's law.

Exam Questions

Q1 a) Two parallel plates are separated by an air gap of 4.5 mm. The plates are connected to a 1500 V dc supply. What is the electric field strength between the plates? Give a suitable unit and state the direction of the field. [3 marks]

b) The plates are now pulled further apart so that the distance between them is doubled. The electric field strength remains the same. What is the new voltage between the plates? [2 marks]

Q2 In an experiment, an alpha particle (charge +2e) was deflected while passing through thin gold foil. The alpha particle passed within 5×10^{-12} m of a gold nucleus (charge +79e). What was the magnitude and direction of the electrostatic force experienced by the alpha particle? (Use $k = 8.99 \times 10^9\,\mathrm{Nm^2C^{-2}}$ and e = -1.6×10^{-19} C.) [4 marks]

Electric fields — one way to roast beef...

At least you get a choice here — uniform or radial, positive or negative, attractive or repulsive, chocolate or strawberry...

Gravitational and Electric Fields

These pages are for Edexcel, AQA A, AQA B, OCR A and OCR B.

Gravitational and electric fields are two quite different things, but many of the equations and diagrams are similar.

There are **Similarities** between **Gravitational** and **Electric Fields**...

1)	Gravitational field strength, *g*, is **force** per **unit mass**.	Electric field strength, *E*, is **force** per **unit positive charge**.
2)	Newton's law of gravitation for the **force** between two point masses is an **inverse square law**. $F \propto \dfrac{1}{r^2}$	Coulomb's law for the electric **force** between two point charges is also an **inverse square law**. $F \propto \dfrac{1}{r^2}$
3)	The **field lines** for a point mass...	The **field lines** for a **negative** point charge...
4)	Gravitational potential, *V*, is **potential energy** per **unit mass** and is **zero** at **infinity**.	Electric potential, *V*, is **potential energy** per **unit positive charge** and is **zero** at **infinity**.

... and some **Differences** too

1) Gravitational forces are always **attractive**. Electric forces can be either **attractive** or **repulsive**.
2) Objects can be **shielded** from **electric** fields, but not from gravitational fields.
3) The size of an **electric** force depends on the **medium** between the charges, e.g. plastic or air. For gravitational forces, this makes no difference.

Energy is Transferred when a Mass or Charge Moves in a Field

1) If a charge in an electric field or a mass in a gravitational field moves **along a field line** then **energy** is **converted** from one form to another.
2) The energy change depends only on where the particle **starts** and **finishes**. It **doesn't matter** what **path** it takes to get from A to B.
3) If the particle moves along an **equipotential**, its **energy doesn't change**.

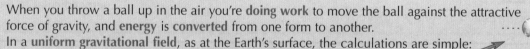

Example in a Gravitational Field

When you throw a ball up in the air you're **doing work** to move the ball against the attractive force of gravity, and **energy** is **converted** from one form to another.
In a **uniform gravitational field**, as at the Earth's surface, the calculations are simple:

1) As the ball rises, it **gains** gravitational **potential energy**: $PE = mg\Delta h$
2) When the ball falls, the gravitational potential energy is converted into **kinetic energy**: $KE = \frac{1}{2}mv^2$ (That's assuming there's no air resistance.)

Example in an Electric Field

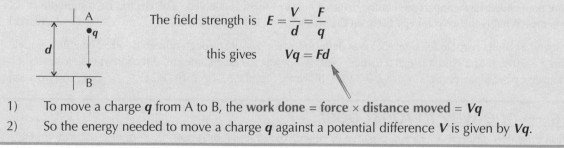

Two parallel plates have a potential difference of *V* across them. This creates a **uniform electric field**.

The field strength is $E = \dfrac{V}{d} = \dfrac{F}{q}$

this gives $Vq = Fd$

1) To move a charge *q* from A to B, the **work done = force × distance moved = Vq**
2) So the energy needed to move a charge *q* against a potential difference *V* is given by *Vq*.

Gravitational and Electric Fields

Absolute Potential is measured Relative to Infinity

1) For both gravitational and electric fields, **potential** is assumed to be **zero** at **infinity** (or where the field strength is so small that it can be neglected).

2) This means that **all objects** will have a potential energy of **zero** at **infinity**.

The next bit is for AQA B only

If Two Masses or Charges Move further apart their Potential Energy Changes

If two particles move closer together or further apart in an electric or gravitational field, their potential energy changes.

1) The **change** in **potential energy** for two **point masses**, m_1 and m_2, that start at a distance r_1 apart, and then move to a distance r_2 apart, is given by:

$$\Delta PE = Gm_1m_2\left(\frac{1}{r_1} - \frac{1}{r_2}\right)$$

Example

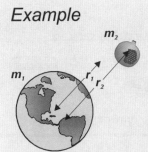

A pomegranate of mass 300 g is catapulted from a height of one metre above the Earth's surface and reaches a height of 20 m above the Earth's surface. The pomegranate gains potential energy as it moves away from the Earth.

mass of the Earth = 6.02×10^{24} kg
radius of the Earth = 6400 km

Its change in potential energy is:

$$\Delta PE = 6.67\times10^{-11}\times6.02\times10^{24}\times0.3\left(\frac{1}{6400001} - \frac{1}{6400020}\right)$$

$$= \underline{\underline{55.9\,J}}$$

2) Similarly the **change** in **potential energy** for two **point charges**, Q_1 and Q_2, is given by

$$\Delta PE = kQ_1Q_2\left(\frac{1}{r_1} - \frac{1}{r_2}\right)$$ where $k = \frac{1}{4\pi\varepsilon}$

3) In a **gravitational** field or an **attractive** electric field — where Q_1 and Q_2 have opposite signs — ΔPE is **positive** when the **distance increases**.

4) In a **repulsive** electric field, ΔPE is **negative** when the distance increases.

Practice Questions

Q1 Describe three similarities and three differences between gravitational and electric fields.

Q2 Define *absolute potential*.

Exam Questions

Q1 Compare the magnitude and direction of the gravitational and electric forces on two electrons which are 8×10^{-10} m apart.
(Use $m_e = 9.11 \times 10^{-31}$ kg, $e = -1.6 \times 10^{-19}$ C, $k = 9.0 \times 10^9$ Nm²C⁻² and $G = 6.67 \times 10^{-11}$ Nm²kg⁻².) [3 marks]

Q2 A negatively charged oil drop is held stationary between two charged plates which are 3 cm apart vertically, and have a potential difference of 5000 V across them.

5000 V

3 cm

(a) The oil drop has mass 1.5×10^{-14} kg. Calculate the size of its charge. [3 marks]
(b) If the polarity of the charged plates were reversed, what would happen to the oil drop? [1 mark]

Save energy — stand on a chair...

Eh? But think about this. When someone says "you've got great potential", they're probably lying. As we all know, gravitational potential is negative at the Earth's surface, and only reaches zero at infinity. Now there's something to aim for.

Capacitors

This page is for Edexcel, AQA A, AQA B, OCR A and OCR B.

Capacitors are things that store electrical charge — like a charge bucket. The capacitance of one of these things tells you how much charge the bucket can hold. Sounds simple enough... ha... ha, ha, ha...

Capacitance *is Defined as the Amount of* Charge Stored *per* Volt

$$C = \frac{Q}{V}$$

where Q is the **charge** in coulombs, V is the **potential difference** in volts and C is the **capacitance** in farads (F) — 1 farad = 1 C V^{-1}.

A farad is a **huge** unit so you'll usually see capacitances expressed in terms of:

μF — microfarads ($\times 10^{-6}$)
nF — nanofarads ($\times 10^{-9}$)
pF — picofarads ($\times 10^{-12}$)

There are Different Types *of* Capacitor

1) There are **several types** of capacitor, most of which you don't have to worry about.

2) The simplest one is made of two metal plates, parallel to each other, separated by an air gap or an insulator called a **dielectric**.

3) In the real world, these plates are two huge sheets of **thin foil** (this makes for a large capacitance — see p. 45) separated by a third sheet of **plastic dielectric**. Then the whole lot's **rolled up** like a swiss roll.

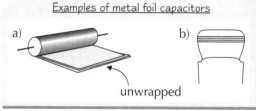

Examples of metal foil capacitors

a)   b)

unwrapped

4) The other type you need to know about is called an **electrolytic capacitor**. The dielectric is just an **oxide film** on the metal foil. They're very **sensitive** to the direction of the voltage, so the positive and negative terminals are clearly marked. ⟶

Examples of electrolytic capacitors

or

> It's really important that you connect these to the correct poles of the battery and that you don't put more than their maximum working voltage through them or they blow up with a VERY BIG BANG!

The **electrical symbol** for a capacitor is: ⊣⊢ or ⊣⊢⁺ ⟵ electrolytic capacitor

You can Investigate *the* Charge Stored *by a* Capacitor Experimentally

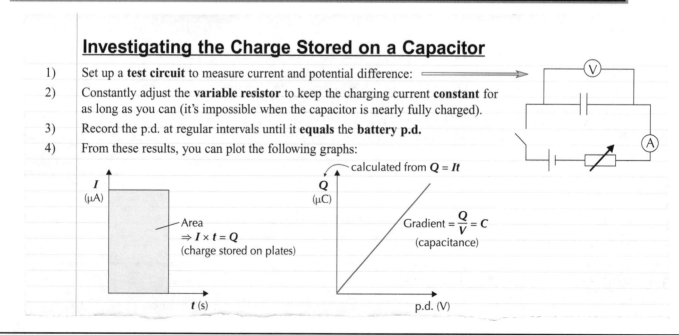

Investigating the Charge Stored on a Capacitor

1) Set up a **test circuit** to measure current and potential difference: ⟹

2) Constantly adjust the **variable resistor** to keep the charging current **constant** for as long as you can (it's impossible when the capacitor is nearly fully charged).

3) Record the p.d. at regular intervals until it **equals** the **battery p.d.**

4) From these results, you can plot the following graphs:

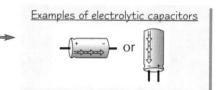

I (μA)

Area
$\Rightarrow I \times t = Q$
(charge stored on plates)

t (s)

calculated from $Q = It$

Q (μC)

Gradient $= \dfrac{Q}{V} = C$
(capacitance)

p.d. (V)

Capacitors

This page is for Edexcel, AQA B and OCR A.

Capacitors in Parallel

1) Adding **capacitors** in **parallel** is like **adding resistors** in **series**.

2) In a **parallel circuit**, the **p.d.** across each component is the **same**.

3) So the **charge** stored on **each** capacitor is: $\boxed{Q = CV}$

4) And the **total charge** stored is: $\boxed{Q_{total} = Q_1 + Q_2 = C_1V + C_2V}$

Total capacitance of capacitors in **parallel** — just **add them up**: $\boxed{C_{total} = C_1 + C_2}$

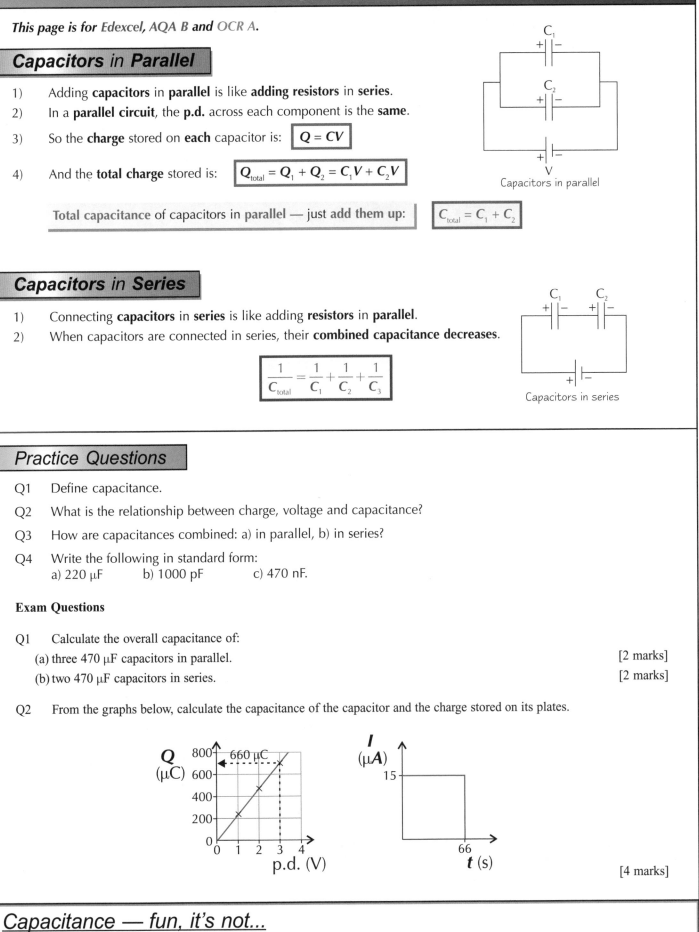

Capacitors in parallel

Capacitors in Series

1) Connecting **capacitors** in **series** is like adding **resistors** in **parallel**.

2) When capacitors are connected in series, their **combined capacitance decreases**.

$$\boxed{\frac{1}{C_{total}} = \frac{1}{C_1} + \frac{1}{C_2} + \frac{1}{C_3}}$$

Capacitors in series

Practice Questions

Q1 Define capacitance.

Q2 What is the relationship between charge, voltage and capacitance?

Q3 How are capacitances combined: a) in parallel, b) in series?

Q4 Write the following in standard form:
a) 220 μF b) 1000 pF c) 470 nF.

Exam Questions

Q1 Calculate the overall capacitance of:
   (a) three 470 μF capacitors in parallel. [2 marks]
   (b) two 470 μF capacitors in series. [2 marks]

Q2 From the graphs below, calculate the capacitance of the capacitor and the charge stored on its plates.

[4 marks]

Capacitance — fun, it's not...

Capacitors are really useful in the real world. Pick an appliance, any appliance, and it'll probably have a capacitor or several. If I'm being honest, though, the only saving grace of this page for me is that it's not especially hard...

Energy Stored & Factors Affecting Capacitance

These pages are for Edexcel, AQA A, AQA B, OCR A and OCR B.

And there's more...

Capacitors *Store Energy*

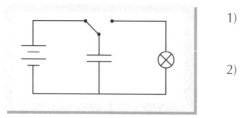

1) In this circuit, when the switch is flicked to the **left**, **charge** builds up on the plates of the **capacitor**. **Electrical energy**, provided by the battery, is **stored** by the capacitor.

2) If the switch is flicked to the **right**, the energy stored on the plates will **discharge** through the **bulb**, converting electrical energy into light and heat.

3) **Work** is done **removing charge** from **one plate** and depositing **opposite charge** onto the other one. The energy for this must come from the **electrical energy** of the **battery**, and is given by **charge × p.d.**

4) So, you can find the **energy stored** by the capacitor from the **area** under a **graph** of **charge stored** against **p.d. across** the capacitor.

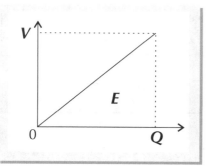

The p.d. across the capacitor is **proportional** to the charge stored on it (see p. 42), so the graph will be a **straight line** through the origin.

The **energy stored** is given by the **yellow triangle**.

5) **Area of triangle = ½ × base × height**, so the energy stored by the capacitor is:

$$E = \frac{1}{2}QV$$

If you're doing AQA A, you need to remember where this equation comes from.

There are *Three* Expressions for the *Energy Stored* by a Capacitor

1) You know the first one already: $E = \dfrac{1}{2}QV$

2) $C = \dfrac{Q}{V}$, so $Q = CV$. Substitute that into the energy equation: $E = \dfrac{1}{2}CV \times V$. So: $E = \dfrac{1}{2}CV^2$

3) $V = \dfrac{Q}{C}$, so $E = \dfrac{1}{2}Q \times \dfrac{Q}{C}$. Simplify: $E = \dfrac{Q^2}{2C}$

Example

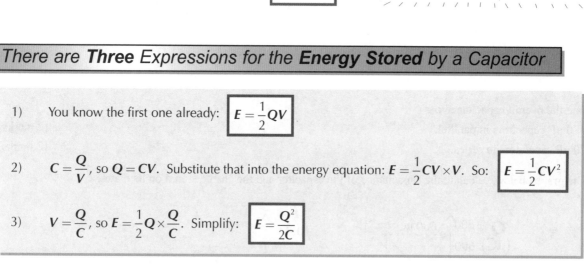

A 900 µF capacitor is charged up to a potential difference of 240 V.
Calculate the energy stored by the capacitor.

First, choose the best equation to use — you've been given **V** and **C**, so you need $E = \dfrac{1}{2}CV^2$.

Substitute the values in: $E = \dfrac{1}{2} \times 9 \times 10^{-4} \times 240 = 0.1$ J

Energy Stored & Factors Affecting Capacitance

This page is for AQA B only.

Several factors affect the capacitance of a parallel plate capacitor — investigate them using the test circuit on p. 42.

Separation of plates: $C \propto \frac{1}{d}$

Capacitance is inversely proportional to the **distance** between the two plates, **d**. To investigate, keep the **charging voltage** and **area** of overlap **constant** and **vary** the separation of plates using spacers. Plot **Q** against $\frac{1}{d}$.

Vary plate separation using spacers:

Area of Overlap of Plates: $C \propto A$

Capacitance is proportional to the **area of overlap** of the two plates (**A**, measured in m²). Keep the **charging voltage** and **separation constant** and **vary** the area of overlap. Plot **Q** against **A**.

Area of overlap **A**

Permittivity of the Dielectric: $C \propto \varepsilon$

The dielectric is the **insulating material** between the plates (e.g. air, paper, mica etc.).
ε (the Greek letter "epsilon") is a constant for the chosen dielectric, called the **absolute permittivity**, measured in Fm⁻¹.

Dielectric: insulating material separating parallel plates.

These variables can be combined into a single equation :

$$C = \frac{\varepsilon A}{d}$$

The absolute permittivity for a material is very small so physicists use **relative permittivity** (ε_r): $\varepsilon_r = \frac{\varepsilon}{\varepsilon_0}$, where ε_0 is the permittivity of free space ($\varepsilon_0 = 8.85 \times 10^{-12}$ Fm⁻¹).

Hence $\varepsilon = \varepsilon_r \varepsilon_0$. So here's an alternative form of the equation:

$$C = \frac{\varepsilon_r \varepsilon_0 A}{d}$$

Practice Questions

Q1   Write down three expressions for the energy stored on a charged capacitor.

Q2   What three factors affect the capacitance of a parallel plate capacitor?

Q3   Draw a circuit diagram of a circuit used to charge up and discharge a capacitor.

Exam Questions

Q1   A 500 μF capacitor is fully charged up from a 12 V supply.

(a) Calculate the total energy stored by the capacitor.   [2 marks]

(b) Calculate the charge stored by the capacitor.   [2 marks]

Q2   A parallel plate capacitor is made from square plates of side 30 cm, separated by a dielectric of relative permittivity 1.8, and with a thickness of 1 mm ($\varepsilon_0 = 8.85 \times 10^{-12}$ Fm⁻¹).

(a) Calculate the capacitance of the capacitor.   [2 marks]

(b) If the length of each side is halved and the thickness of the dielectric is doubled, what effect will this have on the capacitance of the capacitor?   [2 marks]

Chocolate — my kind of energy store...

More equations to learn. Which energy equation you use depends on what you're given in the question... pretty obviously.

Charging and Discharging

These pages are for AQA A, AQA B, OCR A and OCR B only.

Charging and discharging — sounds painful...

You can **Charge** a **Capacitor** by Connecting it to a **Battery**

1) When a capacitor is connected to a **battery**, a **current** flows in the circuit until the capacitor is **fully charged**, then **stops**.

2) The electrons flow onto the plate connected to the **negative terminal** of the battery, so a **negative charge** builds up.

3) This build-up of negative charge **repels** electrons off the plate connected to the **positive terminal** of the battery, making that plate positive. These electrons are attracted to the positive terminal of the battery.

4) An **equal** but **opposite** charge builds up on each plate, causing a **potential difference** between the plates.

    Remember that **no charge** can flow **between** the plates because they're **separated** by an **insulator** (dielectric).

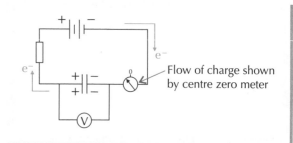
Flow of charge shown by centre zero meter

5) Initially the **current** through the circuit is **high**. But, as **charge** builds up on the plates, **electrostatic repulsion** makes it **harder** and **harder** for more electrons to be deposited. When the p.d. across the **capacitor** is equal to the p.d. across the **battery**, the **current** falls to **zero**. The capacitor is **fully charged**.

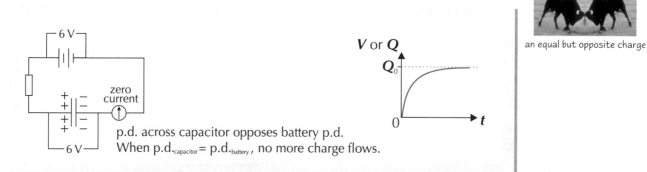

an equal but opposite charge

p.d. across capacitor opposes battery p.d.
When p.d.$_{capacitor}$ = p.d.$_{battery}$, no more charge flows.

To **Discharge** a Capacitor, **Take Out** the **Battery** and **Reconnect** the **Circuit**

1) When a **charged capacitor** is connected across a **resistor**, the p.d. drives a **current** through the circuit.

2) This current flows in the **opposite direction** from the **charging current**.

3) The capacitor is **fully discharged** when the **p.d.** across the plates and the **current** in the circuit are both **zero**.

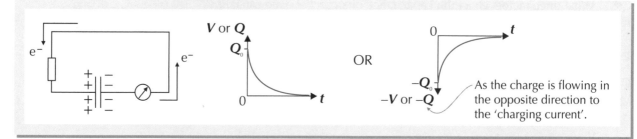

OR

As the charge is flowing in the opposite direction to the 'charging current'.

The **Time Taken** to **Charge** or **Discharge** Depends on **Two Factors**

The **time** it takes to charge up or discharge a capacitor depends on:

1) The **capacitance** of the capacitor (C). This affects the amount of **charge** that can be transferred at a given **voltage**.

2) The **resistance** of the circuit (R). This affects the **current** in the circuit.

Charging and Discharging

The **Charge** on a Capacitor **Decreases Exponentially**

1) When a capacitor is **discharging**, the amount of **charge** left on the plates falls **exponentially with time**.

2) That means it always takes the **same length of time** for the charge to **halve**, no matter **how much charge** you start with — like radioactive decay (see p. 84).

The charge left on the plates of a discharging capacitor is given by the equation:

$$Q = Q_0 e^{-\frac{t}{RC}}$$

where Q_0 is the charge of the capacitor when it's fully charged.

The graphs of V against t and I against t for charging and discharging are also exponential.

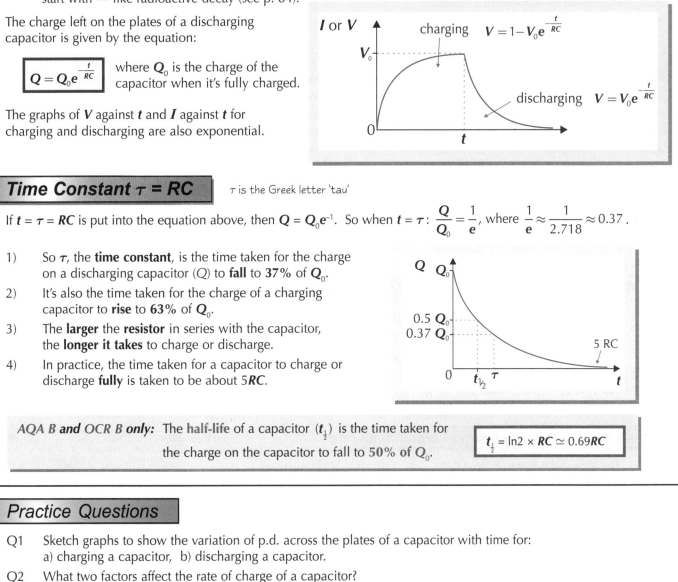

charging $V = 1 - V_0 e^{-\frac{t}{RC}}$

discharging $V = V_0 e^{-\frac{t}{RC}}$

Time Constant τ = RC

τ is the Greek letter 'tau'

If $t = \tau = RC$ is put into the equation above, then $Q = Q_0 e^{-1}$. So when $t = \tau$: $\dfrac{Q}{Q_0} = \dfrac{1}{e}$, where $\dfrac{1}{e} \approx \dfrac{1}{2.718} \approx 0.37$.

1) So τ, the **time constant**, is the time taken for the charge on a discharging capacitor (Q) to **fall** to **37%** of Q_0.

2) It's also the time taken for the charge of a charging capacitor to **rise** to **63%** of Q_0.

3) The **larger** the **resistor** in series with the capacitor, the **longer it takes** to charge or discharge.

4) In practice, the time taken for a capacitor to charge or discharge **fully** is taken to be about 5RC.

AQA B and OCR B only: The **half-life** of a capacitor ($t_{\frac{1}{2}}$) is the time taken for the charge on the capacitor to fall to **50%** of Q_0.

$$t_{\frac{1}{2}} = \ln 2 \times RC \simeq 0.69RC$$

Practice Questions

Q1 Sketch graphs to show the variation of p.d. across the plates of a capacitor with time for:
a) charging a capacitor, b) discharging a capacitor.

Q2 What two factors affect the rate of charge of a capacitor?

Exam Questions

Q1 A 250 μF capacitor is fully charged from a 6 V battery and then discharged through a 1 kΩ resistor.

(a) Calculate the time taken for the charge on the capacitor to fall to 37% of its original value. [2 marks]

(b) Calculate the time taken for the charge on the capacitor to fall to 50% of its original value. *(AQA B & OCR B)* [2 marks]

(c) Calculate the percentage of the total charge remaining on the capacitor after 0.7s. [3 marks]

(d) If the charging voltage is increased to 12 V, what effect will this have on:

i) the total charge stored

ii) the capacitance of the capacitor

iii) the time taken to fully charge [3 marks]

An analogy — consider the lowly bike pump...

A good way to think of the charging process is like pumping air into a bike tyre. To start with, the air goes in easily, but as the pressure in the tyre increases, it gets harder and harder to squeeze any more air in. The tyre's 'full' when the pressure of the air in the tyre equals the pressure of the pump. The analogy works just as well for discharging...

Magnetic Fields

This page is for Edexcel, AQA A, AQA B, OCR A and OCR B.

A **Magnetic Field** is a **Region** Where a **Force** is Exerted on **Magnetic Materials**

1) Magnetic fields can be represented by **field lines**.
2) Field lines go from **north to south**.
3) The **closer** together the lines, the **stronger** the field.

At a neutral point magnetic fields cancel out.

There is a **Magnetic Field** Around a **Wire** Carrying **Electric Current**

1) The **direction** of a magnetic **field** around a current-carrying wire can be worked out with the **right-hand rule**.
2) You also need to learn these diagrams for a **single coil** and a **solenoid**.

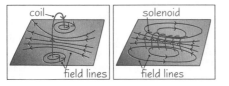

RIGHT-HAND RULE

1) Stick your right thumb up, like you're hitching a lift.
2) If your thumb points in the direction of the current...
3) ...your curled fingers point in the direction of the field.

A **Wire** Carrying a **Current** in a **Magnetic Field** will **Experience** a **Force**

1) A **force acts** if a current-carrying wire **cuts magnetic flux** lines.
2) If the current is **parallel** to the flux lines, **no force** acts.
3) The **direction** of the force is always **perpendicular** to both the **current** direction and the **magnetic field**.
4) The direction of the force is given by **Fleming's Left-Hand Rule**.

Fleming's Left-Hand Rule
The First finger points in the direction of the uniform magnetic Field, the seCond finger points in the direction of the conventional Current. Then your thuMb points in the direction of the force (in which Motion takes place).

The **Size** of the **Force** can be **Calculated...**

1) The size of the **force**, F, on a current-carrying wire at right-angles to a magnetic field is proportional to the **current**, I, the **length of wire** in the field, l, and the **strength of the magnetic field**, B. This gives the equation:

$$F = BIl$$

2) In this equation, the **magnetic field strength**, B, is defined as:

The size of the **force per metre** of length per unit of **current** flowing in a wire at **right angles** to a **magnetic field**.

3) **Magnetic field strength** is also called **flux density** and it's measured in **tesla, T**.
4) Magnetic field strength is a **vector** quantity with both a **direction** and **magnitude**.

It helps to think of flux density as the number of flux lines per unit area.

The Force is **Greatest** when the **Wire** and **Field** are **Perpendicular...** *OCR A only*

1) The **force** on a current-carrying wire in a magnetic field is caused by the **component** of field strength which is **perpendicular** to the wire — $B\sin\theta$.
2) So, for a wire at an **angle** θ to the field, the **force** acting on the wire is given by:

$$F = BIl\sin\theta$$

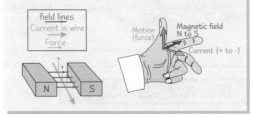

Examples:

→ current
→ magnetic field

i) If $\theta = 90°$, $F = BIl$
ii) If $\theta = 30°$, $F = BIl×0.5$
iii) If $\theta = 0°$, $F = 0$

Magnetic Fields

This page is for Edexcel, AQA B and OCR B only.

A Hall Probe is Used to Measure Field Strength   *Edexcel only*

1) A **Hall probe** is a device that creates a **potential difference** when placed in a **magnetic field**.

2) This potential difference can be **measured**, and is **proportional** to the **field strength**, *B*.

3) This means that a **Hall probe** can be used to **measure** magnetic **field strength** — so long as it's been **pre-calibrated** against a **known value** of *B* first.

4) The field strengths around a **straight wire** and at the centre of a **solenoid** can be **investigated** using a Hall probe.

5) **Experimental** results have been used to find the following **equations** for magnetic field strength:

Near a **straight wire**: $B = \dfrac{\mu_0 I}{2\pi r}$

At the **centre** of a **long solenoid**: $B = \mu_0 n I$

Where I = current, r = distance from wire, μ_0 = ("mu-nought") permeability of free space, and n = number of turns in the solenoid.

A Current Balance Measures the Force on a Wire in a Magnetic Field   *AQA B only*

1) A **current balance** consists of a **rectangular loop** of wire **balanced** on a **pivot**, with **one side** of the loop in a magnetic **field**.

2) When a **current** is passed through the wire a downward **force** acts on the loop (use Fleming's left-hand rule to check this out for yourself).

3) **Mass** can be added to a tray on the other side until the **forces balance** — this happens when *mg = BIl*.

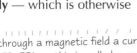

F = mg   pivot   F = BIL

The Forces on a Loop Can Be Used To Make A Motor   *OCR B and AQA B only*

Force

axis

+ve

−ve

Split-ring commutator   Force

1) If a **current-carrying loop** is placed in a **magnetic field**, the **forces** on the side arms cause the loop to **rotate**.

2) By using a **split-ring commutator**, the current in the loop can be **reversed** each time the loop becomes **vertical** (i.e. every **half turn**).

3) This allows the loop to **rotate steadily** — which is otherwise known as a **motor**.

If a wire loop is moved through a magnetic field a current is induced in it (see pages 52–53) — this is called a generator.

Practice Questions

Q1 Sketch the magnetic fields around a long straight current-carrying wire, and a solenoid. Show the direction of the current and magnetic field on each diagram.

Q2 A copper bar can roll freely on two copper supports, as shown in the diagram. When current is applied in the direction shown, which way will the bar roll?

Horseshoe Magnet

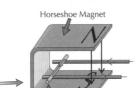
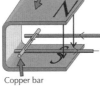

Copper bar

Exam Questions

Q1 A wire carrying a current of 3 A runs perpendicular to a magnetic field of strength 2×10^{-5} T. 4 cm of the wire is within the field.

(a) Calculate the magnitude of the force on the wire. [2 marks]

(b) If the wire is rotated so that it is at 30° to the field, what would the size of the force be? *(OCR A only)* [2 marks]

Magnetic fields — great for iron cows...

Fleming's left hand rule is the key to this section — so make sure you know how to use it and understand what it all means. Remember that the direction of magnetic field is from N to S, and that the current is from +ve to −ve — this is as important as using the correct hand, otherwise it'll all go to pot...

Motion of Charged Particles in a B Field

This page is for AQA A, AQA B, OCR A and OCR B only.

Magnetic fields are used a lot when dealing with particle beams — you'll be learning more about their uses in Section 9.

Forces Act on Charged Particles in Magnetic Fields

Electric current in a wire is caused by the **flow** of negatively **charged** electrons. These charged particles are affected by **magnetic fields** — so a current-carrying wire **moves** in a magnetic field (see pages 48–49).

1) The equation for the **force** exerted on a **current-carrying wire** in a **magnetic field** perpendicular to the current is:

   Equation 1: $\boxed{F = BIl}$

2) To see how this relates to **charged particles** moving through a wire, you need to know that electric **current**, I, is the flow of **charge**, q, per unit **time**, t.

   $\boxed{I = \dfrac{q}{t}}$

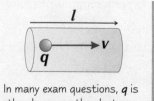

In many exam questions, q is the charge on the electron, which is 1.6×10^{-19} coulomb.

3) A charged particle which moves a **distance** l in **time** t has a **velocity**, v:

   $\boxed{v = \dfrac{l}{t} \Rightarrow t = \dfrac{l}{v}}$

4) So, putting the two equations **together** gives the **current** in terms of the **charge** flowing through the **wire**:

   Equation 2: $\boxed{I = \dfrac{qv}{l}}$

5) Put **equation 2** back into **equation 1** to give the **electromagnetic force** on the wire as:

   $\boxed{F = Bqv}$

6) You can use this equation to find the **force** acting on a **single charged particle** moving through a magnetic field.

Example
What is the force acting on an electron travelling at $2 \times 10^4 \text{ ms}^{-1}$ through a uniform magnetic field of strength 2 T? (The charge on an electron is 1.6×10^{-19} C.)

$$F = Bqv$$
so, $\quad F = 2 \times 1.6 \times 10^{-19} \times 2 \times 10^4$
so, $\quad F = 6.4 \times 10^{-15}$ N

Charged Particles in a Magnetic Field are Deflected in a Circular Path

1) By **Fleming's left hand rule** the force on a **moving charge** in a magnetic field is always **perpendicular** to its **direction of travel**.

2) Mathematically, that is the condition for **circular** motion.

3) This effect is used in **particle accelerators** such as **cyclotrons** and **synchrotrons** (see pages 102–103) — which use **magnetic fields** to accelerate particles to very **high energies** along circular paths .

4) The **radius of curvature** of the **path** of a charged particle moving through a magnetic field gives you information about the particle's **charge** and **mass** — this means you can **identify different particles** by studying how they're **deflected** (see pages 104–105).

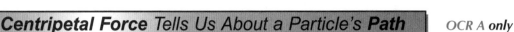

Centripetal Force Tells Us About a Particle's Path   *OCR A only*

The centripetal force and the electromagnetic force are equivalent for a charged particle travelling along a circular path.

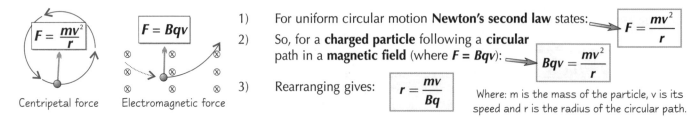

1) For uniform circular motion **Newton's second law** states: $\boxed{F = \dfrac{mv^2}{r}}$

2) So, for a **charged particle** following a **circular** path in a **magnetic field** (where $F = Bqv$): $\boxed{Bqv = \dfrac{mv^2}{r}}$

3) Rearranging gives: $\boxed{r = \dfrac{mv}{Bq}}$

Where: m is the mass of the particle, v is its speed and r is the radius of the circular path.

Motion of Charged Particles in a B Field

This page is for AQA B only.

Specific **Particle Velocities** can be **Selected** for Experiments

1) A positively **charged particle** travelling through an **electric field** between two charged plates will be attracted towards the negative plate with a **force**, **F = Eq** (where *E* is the electric field strength, measured in NC⁻¹, and q is the charge on the particle).

2) As you've already seen, a charged particle travelling through a **magnetic field** will also be deflected by a **force**, **F = Bqv**.

3) So, a **charged particle** that passes through an **electric** and a **magnetic field** will have **forces acting on it** due to **both** of these fields.

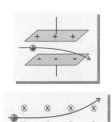

4) The diagram below shows a **device** that uses this effect to **control the paths** of charged particles — this lets you **select** particles of a **specific velocity**, which is why the device is called a **velocity selector**.

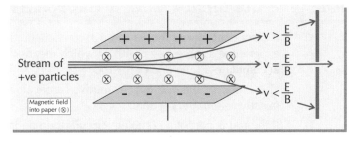

5) In a **velocity selector** an **electric field** (**E**) and a **magnetic field** (**B**) are set up to be **perpendicular** to each other and to the direction of a beam of charged particles.

6) When the **forces** due to E and B are **balanced**, the particles will **not be deflected** as they travel through the velocity selector — this happens when: $\boxed{Bqv = Eq}$

7) Rearranging this gives the **velocity**, *v*, of the particles that experience **balanced forces**: $\boxed{v = \dfrac{E}{B}}$

8) Particles with a **higher or lower velocity** than this will be **deflected**, and only particles with a velocity **equal** to *E/B* will remain **undeflected**.

9) By using a **screen** with a **hole** in it, called a **collimator plate**, deflected particles can be **filtered out**, leaving a beam of **undeflected particles** which all have the **same velocity**.

Practice Questions

Q1 Derive the formula for the force on a charged particle in a magnetic field, **F = Bqv**, from **F = BIl**.

Q2 Explain why a particle with a velocity $v = \dfrac{E}{B}$ remains undeflected in a velocity selector.

Exam Questions

Q1 (a) What is the force on an electron travelling at a velocity of 5×10^6 ms⁻¹ through a perpendicular magnetic field of 0.77 T? [The charge on an electron is -1.6×10^{-19} C.] [2 marks]

   (b) Explain why it follows a circular path while in the field. [1 mark]

Q2 What is the radius of the circular path of an electron with a velocity of 2.3×10^7 ms⁻¹ moving perpendicular to a magnetic field of 0.6 mT? *(OCR A only)*
[The mass of an electron is 9.11×10^{-31} kg and its charge is -1.6×10^{-19} C.] [3 marks]

Q3 A velocity selector uses an electric field of 3.75×10^4 NC⁻¹ and a uniform magnetic field of 1.5 mT, at right angles to the electric field. A charged particle travels undeflected through the velocity selector, perpendicular to both fields. What velocity is the particle travelling at? [2 marks]

Hold on to your hats folks, this is starting to get tricky...

Basically, the main thing you need to know here is that both electric and magnetic fields will exert a force on a charged particle, making it follow a circular path. If these two forces are balanced, the particle won't be deflected — which is how a velocity selector works. Now all you need to do is learn the equations...

Electromagnetic Induction

These pages are for Edexcel, AQA A, AQA B, OCR A and OCR B.

Think of the **Magnetic Flux** as the Total **Number** of **Field Lines**...

1) Magnetic field strength, or **magnetic flux density**, *B*, is a measure of the **strength** of the magnetic field **per unit area**.

2) So, the total **magnetic flux**, φ, passing through an **area**, *A*, perpendicular to a **magnetic field**, *B*, is defined as:
$$\phi = BA$$

3) For a **solenoid**, you measure the **flux linkage**, Φ, which for a coil of *N* turns is given by: $\Phi = N\phi = BAN$

φ is the little Greek letter 'phi', and Φ is a capital 'phi'.
(The unit of both φ and Φ is the weber, Wb. 1 tesla = 1 Wb m^{-2}.)

Example

Area, $A = 3 \text{ m}^2$

Flux density, $B = 4 \times 10^{-3}$ T
(flux per unit area)

$\phi = BA$
$= 4 \times 10^{-3} \times 3$
$= 12 \times 10^{-3}$ Wb

If this is the magnetic flux inside a solenoid of 10 turns, the flux linkage will be $\Phi = N\phi = 0.12$ Wb

Charges Accumulate on a Conductor Moving Through a Magnetic Field

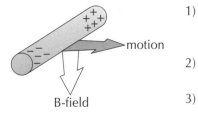

motion

B-field

1) If a **conducting rod** moves through a magnetic field its **electrons** will experience a **force** (see p. 48) — which means that they will **accumulate** at one end of the rod.

2) This **induces** an **e.m.f.** (**electromotive force**) across the ends of the rod exactly as a **battery** would.

3) If the rod is part of a complete **circuit**, then an induced **current** will **flow** through it — this is called **electromagnetic induction**.

Changes in Magnetic Flux Induce an Electromotive Force

1) An **electromotive force** (**e.m.f.**) is **induced** when there is **relative motion** between a **conductor** and a **magnet**.

2) The **conductor** can **move** and the **magnetic field** stay **still** or the **other way round** — you get an e.m.f. either way.

3) An **e.m.f.** is **produced** whenever **lines of force** (flux) are **cut**.

4) **Flux cutting** always induces e.m.f. but will only **induce** a **current** if the **circuit** is complete.

5) **Flux linking** is when an e.m.f. is induced by **changing** the **magnitude** or **direction** of the magnetic flux (e.g. caused by an **alternating current** electromagnet).

Coil 1 Coil 2

For example, if the **voltage** in Coil 1 is **increased**, an **e.m.f.** and a **current** will be **induced** in Coil 2 — this is the principle used in **transformers** (see p.54).

These Results are Summed up by Faraday's Law...

FARADAY'S LAW: The **induced e.m.f.** is **directly proportional** to the **rate of change of flux linkage**.

1) **Faraday's Law** can be written as: Induced e.m.f. $= \dfrac{\text{flux change}}{\text{time taken}} = \dfrac{\Delta\Phi}{\Delta t}$

2) The **size** of the e.m.f. is shown by the **gradient** of a graph of Φ against time.

3) The **area under** the graph of e.m.f. against time gives the **flux change**.

Φ e.m.f. = gradient

time

Φ = area

time

Exam questions often ask you to calculate the e.m.f. induced by the Earth's magnetic field across the wingspan of a plane. Think of it as a moving rod and use the equation above.

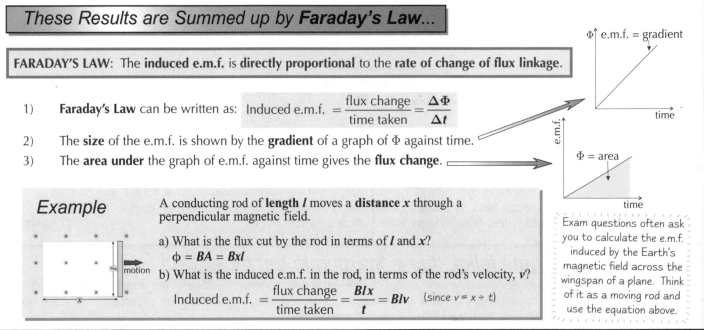

Example

A conducting rod of **length** *l* moves a **distance** *x* through a perpendicular magnetic field.

a) What is the flux cut by the rod in terms of *l* and *x*?
$$\phi = BA = Bxl$$

b) What is the induced e.m.f. in the rod, in terms of the rod's velocity, *v*?

Induced e.m.f. $= \dfrac{\text{flux change}}{\text{time taken}} = \dfrac{Blx}{t} = Blv$ (since $v = x \div t$)

motion

Electromagnetic Induction

The **Direction** of the **Induced E.m.f.** and **Current** are given by **Lenz's Law...**

LENZ'S LAW: The induced e.m.f. is always in such a direction as to oppose the change that caused it.

1) **Lenz's law** and **Faraday's law** can be **combined** to give one formula that works for both:

$$\text{Induced e.m.f.} = -\frac{d\Phi}{dt}$$

2) The **minus sign** shows the direction of the **induced e.m.f.**

3) The idea that an induced e.m.f. will **oppose** the change that caused it agrees with the principle of the **conservation of energy** — the **energy used** to pull a conductor through a magnetic field, against the **resistance** caused by magnetic **attraction**, is what **produces** the **induced current**.

4) **Lenz's law** can be used to find the **direction** of an **induced e.m.f.** and **current** in a conductor travelling at right angles to a magnetic field...

1) **Lenz's law** says that the **induced e.m.f.** will produce a force that **opposes** the motion of the conductor — in other words a **resistance**.

2) Using **Fleming's left-hand rule** (see p.48), point your thumb in the direction of the force of **resistance** — which is in the **opposite direction** to the motion of the conductor.

3) Your **second finger** will now give you the direction of the **induced e.m.f.**

4) If the conductor is **connected** as part of a **circuit**, a current will be induced in the **same direction** as the induced e.m.f.

Practice Questions

Q1 What is the difference between magnetic flux density, magnetic flux and magnetic flux linkage?

Q2 State Faraday's law.

Q3 State Lenz's law.

Q4 A coil consists of N turns each of area A. If it is placed at right angles to a uniform magnetic field, what is its flux linkage?

Q5 Explain how you can find the direction of an induced e.m.f. in a copper bar moving at right angles to a magnetic field.

Exam Questions

Q1 A coil of area 0.23 m² is placed at right angles to a magnetic field of 2×10^{-3} T.
  (a) What is the magnetic flux passing through the coil? [2 marks]
  (b) If the coil has 150 turns, what is the magnetic flux linkage in the coil? [2 marks]
  (c) Over a period of 2.5 seconds the magnetic field is reduced uniformly to 1.5×10^{-3} T. What is the size of the e.m.f. induced across the ends of the coil? [3 marks]

Q2 An aeroplane with a wingspan of 30 m flies at a speed of 100 ms⁻¹ perpendicular to the Earth's magnetic field. The Earth's magnetic field at the aeroplane's location is 60×10^{-6} T.
  (a) Calculate the induced e.m.f. between the wing tips. [2 marks]
  (b) Complete the diagram to show the direction of the induced e.m.f. between the wing-tips. [1 mark]

Beware — physics can induce extreme confusion...

Make sure you know the difference between flux and flux linkage, and that you can calculate both. Then all you need to learn is that the induced e.m.f. is proportional to minus the rate of change of flux linkage — and that's it. Remember when you're using Fleming's left-hand rule to work out the direction of the induced e.m.f. that you need to point your thumb in the opposite direction to the direction the conductor is moving in...

Transformers and Alternators

This page is for Edexcel, AQA B and OCR B only.

Transformers Work by Electromagnetic Induction

1) **Transformers** are devices that make use of electromagnetic induction to **change** the size of the **voltage** for an **alternating current**.

2) An alternating current flowing in the **primary** (or input) **coil** produces **magnetic flux**.

3) The **magnetic field** is passed through the **iron core** to the **secondary** (or output) coil, where it **induces** an alternating **voltage** of the same frequency.

4) From Faraday's law , the **induced** e.m.f.'s in both the **primary** and **secondary** coils can be calculated:

Primary coil

$$V_p = N_p \frac{d\Phi}{dt}$$

Secondary coil

$$V_s = N_s \frac{d\Phi}{dt}$$

These can be combined to give the equation for an **ideal transformer**:

$$\frac{V_p}{V_s} = \frac{N_p}{N_s}$$

(where N is the number of turns in a coil)

5) **Step-up** transformers **increase** the **voltage** by having **more turns** on the **secondary** coil than the primary. **Step-down** transformers **reduce** the voltage by having **fewer** turns on the secondary coil.

> *Example* What is the output voltage for a transformer with a primary coil of 100 turns, a secondary coil of 300 turns and an input voltage of 230 V?
>
> $$\frac{V_p}{V_s} = \frac{N_p}{N_s} \quad \Rightarrow \quad \frac{230}{V_s} = \frac{100}{300} \quad \Rightarrow \quad V_s = \frac{230 \times 300}{100} = 690 \text{ V}$$

Transformers are Not 100% Efficient

1) If a transformer was **100% efficient** the **power in** would **equal** the **power out**.

2) This means that for an **ideal transformer**: $\boxed{V_p I_p = V_s I_s}$ or $\boxed{\dfrac{V_p}{V_s} = \dfrac{I_s}{I_p}}$

You can put the two ideal transformer equations together to give:

$$\frac{V_p}{V_s} = \frac{N_p}{N_s} = \frac{I_s}{I_p}$$

3) However, in practice there will be **small losses** of **power** from the transformer, mostly in the form of **heat**.

4) **Heat** can be produced by **eddy currents** in the transformer's iron **core** — this effect is reduced by **laminating** the core with layers of **insulation**.

5) Heat is also generated by **resistance** in the coils — to minimise this, **thick copper wire** is used, which has a **low resistance**.

6) The **efficiency** of a transformer is simply the **ratio** of **power out** to **power in**, so: $\Rightarrow$ $\boxed{\text{efficiency} = \dfrac{V_s I_s}{V_p I_p}}$

Transformers are an Important Part of the National Grid...

1) **Electricity** from power stations is sent round the country in the **National Grid** at the **lowest** possible current, because **losses** due to the **resistance** in the cables are proportional to I^2 — so if you double the transmitted current, you quadruple the power lost.

2) Since **power = current × voltage**, a **low current** means a **high voltage**.

3) **Transformers** allow us to **step up** the voltage to around **400 000 V** for **transmission** through the national grid, and then **reduce** it again to **230 V** for domestic use.

400 kV

25 kV

230 V

Power station

Step-up transformer

Step-down transformer

Home

... robots in disguise

Transformers and Alternators

This page is for AQA B and OCR B only.

An *Alternator* is a *Generator* of *Alternating Current*

1) **Generators**, or dynamos, **convert** kinetic energy into **electrical energy** — they **induce** an electric **current** by **rotating** a **coil** in a magnetic field.

2) A simple **alternator** looks similar to a **motor** but with **slip rings** and **brushes** instead of split-ring commutators.

3) The output **voltage** and **current** change direction with every **half rotation** of the coil, producing **alternating current** (**AC**).

Flux Linkage and *Induced Voltage* are *90° Out of Phase*   *OCR B only*

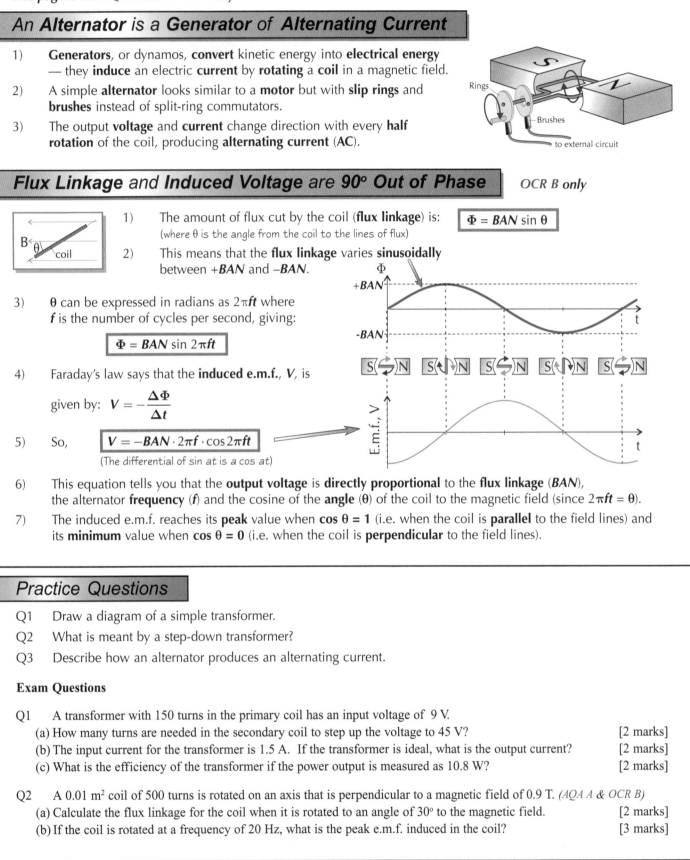

1) The amount of flux cut by the coil (**flux linkage**) is: $\Phi = BAN \sin \theta$
   (where θ is the angle from the coil to the lines of flux)

2) This means that the **flux linkage** varies **sinusoidally** between +**BAN** and –**BAN**.

3) θ can be expressed in radians as $2\pi ft$ where *f* is the number of cycles per second, giving:

$$\Phi = BAN \sin 2\pi ft$$

4) Faraday's law says that the **induced e.m.f.**, *V*, is given by: $V = -\dfrac{\Delta\Phi}{\Delta t}$

5) So, $\boxed{V = -BAN \cdot 2\pi f \cdot \cos 2\pi ft}$
   (The differential of sin *at* is a cos *at*)

6) This equation tells you that the **output voltage** is **directly proportional** to the flux linkage (**BAN**), the alternator **frequency** (*f*) and the cosine of the **angle** (θ) of the coil to the magnetic field (since $2\pi ft = \theta$).

7) The induced e.m.f. reaches its **peak** value when **cos θ = 1** (i.e. when the coil is **parallel** to the field lines) and its **minimum** value when **cos θ = 0** (i.e. when the coil is **perpendicular** to the field lines).

Practice Questions

Q1 Draw a diagram of a simple transformer.

Q2 What is meant by a step-down transformer?

Q3 Describe how an alternator produces an alternating current.

Exam Questions

Q1 A transformer with 150 turns in the primary coil has an input voltage of 9 V.
  (a) How many turns are needed in the secondary coil to step up the voltage to 45 V?   [2 marks]
  (b) The input current for the transformer is 1.5 A. If the transformer is ideal, what is the output current?   [2 marks]
  (c) What is the efficiency of the transformer if the power output is measured as 10.8 W?   [2 marks]

Q2 A 0.01 m² coil of 500 turns is rotated on an axis that is perpendicular to a magnetic field of 0.9 T. *(AQA A & OCR B)*
  (a) Calculate the flux linkage for the coil when it is rotated to an angle of 30° to the magnetic field.   [2 marks]
  (b) If the coil is rotated at a frequency of 20 Hz, what is the peak e.m.f. induced in the coil?   [3 marks]

Arrrrrrrrrggggggggghhhhhhhh...

Breathe a sigh of relief, pat yourself on the back and make a brew — well done, you've reached the end of Section Six. That was pretty nasty stuff, but don't let all of those equations get you down — once you've learnt the main ones and can use them blindfolded, even the trickiest looking exam question will be a walk in the park...

Charge/Mass Ratio of the Electron

*These pages are for **Edexcel** (just the electron-volt bit), **AQA A Option 8**, **AQA B** and **OCR B** only.*

e/m was known for quite a long time before anyone came up with a way to measure one or the other separately.

Cathode Ray *is an* Old-Fashioned *name for the* Electron

1) The phrase '**cathode ray**' was first used in 1876, to describe the **glow** that appears on the wall of a discharge tube like the one in the diagram, when a **potential difference** is applied across the terminals.

2) The **rays** seemed to come from the **cathode** (hence their name) and there was a lot of argument about **what** the rays were made of.

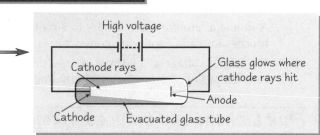

3) **J. J. Thomson** ended the debate in 1897, when he demonstrated (see opposite) that cathode rays:

   a) have **energy**, **momentum** and **mass**,

   b) have a **negative charge**,

   c) have the **same properties**, no matter **what gas** is in the tube and what the **cathode** is made of,

   d) have a **charge to mass ratio** much **bigger** than that of **hydrogen** ions. So they either have a **tiny mass**, or a much higher charge than protons — Thomson assumed they had the same size charge as protons.

Thomson concluded that **all atoms** contain these 'cathode ray particles', that were later called **electrons**.

Electron Beams *are Produced by* Thermionic Emission

1) When you **heat** a **metal**, its **free electrons** gain a load of **thermal energy**.

2) Give them **enough energy** and they'll **break free** from the surface of the metal — this is called **thermionic emission**. (Try breaking the word down — think of it as '**therm**' [to do with heat] + '**ionic**' [to do with charge] + '**emission**' [giving off] — so it's 'giving off charged particles when you heat something'.)

3) Once they've been emitted, the electrons can be **accelerated** by an **electric field** in an **electron gun**:

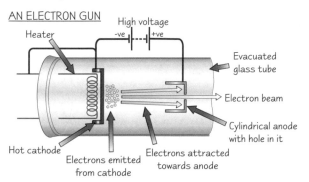

A **heating coil** heats the metal cathode. The electrons that are emitted are **accelerated** towards the **cylindrical anode** by the electric field set up by the high voltage.

Some electrons pass through a **little hole** in the **anode**, making a narrow electron beam. The electrons in the beam move at a **constant velocity** because there's **no field** beyond the anode — i.e., there's **no force**.

The Electron-Volt *is Defined Using* Accelerated Charges

1) The **kinetic energy** that a particle with charge **Q gains** when it's **accelerated** through a p.d. of **V** volts is **QV** joules. That just comes from the definition of the **volt** (JC^{-1}).

2) If you replace **Q** in the equation with the charge of a **single electron**, **e**, you get: $\Rightarrow$

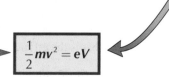

$$\frac{1}{2}mv^2 = eV$$

3) From this you can define a new **unit of energy** called the **electron-volt** (**eV**):

> 1 electron-volt is the **kinetic energy carried** by an **electron** after it has been **accelerated** through a **potential difference** of **1 volt**.

4) So, the **energy in eV** of an electron accelerated by a potential difference is:

energy gained by electron (eV)
= accelerating voltage (V)

Conversion factor: 1 eV = 1.6 × 10^{-19} J

Charge/Mass Ratio of the Electron

Thomson *Measured the* Specific Charge *of the* Electron

1) The **specific charge** or **charge/mass ratio** of a charged particle is just its **charge** per unit **mass**.

2) There are a **few different ways** of measuring it, and you need to know about **one** of them.
This isn't the method that Thomson used, but that's not important.

3) All the **physics** behind this experiment is covered in *Section 6 — Electromagnetism*.

Measuring the Charge/Mass Ratio of an Electron

1) Electrons are charged particles, so they can be deflected by an **electric** or a **magnetic field**. This method uses a magnetic field in a piece of apparatus called a **fine beam tube**.

2) When the beam of electrons from the **electron gun** (see previous page) passes through the low-pressure gas, the hydrogen atoms along its path **absorb energy**. As the electrons in these **excited hydrogen atoms** fall back to the ground state, they **emit light** (see p. 66). The electron beam is seen as a **glowing trace** through the gas.

magnetic field coils

electron gun

electron beam

glass bulb containing
hydrogen at low pressure

3) Two circular **magnetic field coils** either side generate a **uniform magnetic field** inside the tube.

4) The electron beam is initially fired at **right angles** to the **magnetic field**, so the beam curves round in a **circle**.

5) This means that the **magnetic force** on the electron (see p. 50) is acting as a **centripetal force** (see p. 13). So the radius of the circle is given by:

$$\frac{mv^2}{r} = Bev$$

where *m* is the mass of an electron, *e* is the charge on an electron, *B* is the magnetic field strength, *v* is the velocity of the electron and *r* is the radius of the circle.

6) From the previous page, you've got an equation that you can rearrange to give *v* in terms of the **accelerating potential** of the electron gun. If you substitute that expression for *v* into the equation above (and tidy it all up a bit) you get:

$$\frac{e}{m} = \frac{2V}{B^2 r^2}$$

where *m* is the mass of an electron, *e* is the charge on an electron, *B* is the magnetic field strength, *V* is the accelerating potential and *r* is the radius of the circle.

You can **measure** all the quantities on the **right-hand side** of the equation using the **fine beam tube**, so you're left with the **specific charge**, *e*/*m*.

Practice Questions

Q1 What is meant by thermionic emission?

Q2 Sketch a labelled diagram of an electron gun, which could be used to accelerate electrons.

Q3 What was Thomson's main conclusion following his measurement of e/m for electrons?

Exam Questions

Q1 An electron of mass 9.1×10^{-31} kg and charge -1.6×10^{-19} C is accelerated through a potential difference of 1 kV.

  (a) Write down its energy in eV.   [1 mark]

  (b) Calculate its energy in joules.   [1 mark]

  (c) Calculate its speed in ms^{-1} and express this as a percentage of the speed of light (3.0×10^8 ms^{-1}).   [3 marks]

Q2 Explain the main features of an experiment to determine the specific charge of the electron.
You may be awarded a mark for the clarity of your answer.   [5 marks]

New Olympic event — the electron-vault...

Electron-volts are really handy units — they'll crop up all over the rest of this book, particularly in nuclear and particle physics. It stops you having to mess around with a load of nasty powers of ten. Cathode ray tubes (CRTs) are pretty handy too — there's one in your telly... unless you've got one of those new-fangled flat-screen plasma thingummy-do-dahs...

Millikan's Oil-Drop Experiment

These pages are for AQA A Option 8 and OCR B only.

Thomson had already found the charge/mass ratio of the electron in 1897 — now it was down to Robert Millikan, experimenter extraordinaire, to find the absolute charge...

Millikan's Experiment used Stoke's Law

1) Before you start thinking about Millikan's experiment, you need a bit of **extra theory**.

2) When you drop an object into a fluid, like air, it experiences a **viscous drag** force. This force acts in the **opposite direction** to the velocity of the object, and is due to the **viscosity** of the fluid.

3) You can calculate this viscous force on a spherical object using **Stoke's law**:

$$F = 6\pi\eta rv$$

where η is the viscosity of the fluid, r is the radius of the object and v is the velocity of the object.

Millikan's Experiment — the Basic Set-Up

Millikan's Oil-Drop Experiment — Apparatus

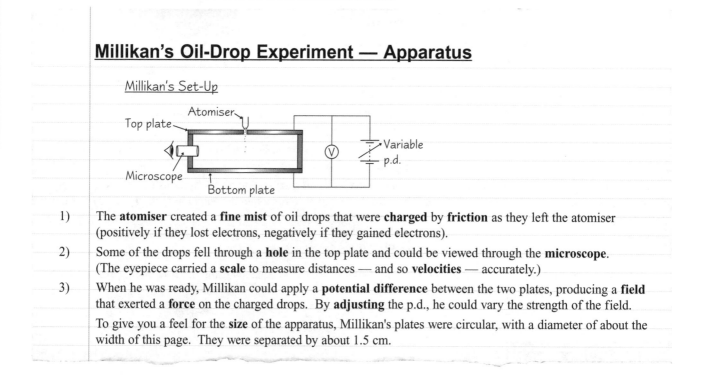

Millikan's Set-Up

1) The **atomiser** created a **fine mist** of oil drops that were **charged** by **friction** as they left the atomiser (positively if they lost electrons, negatively if they gained electrons).

2) Some of the drops fell through a **hole** in the top plate and could be viewed through the **microscope**. (The eyepiece carried a **scale** to measure distances — and so **velocities** — accurately.)

3) When he was ready, Millikan could apply a **potential difference** between the two plates, producing a **field** that exerted a **force** on the charged drops. By **adjusting** the p.d., he could vary the strength of the field.

To give you a feel for the **size** of the apparatus, Millikan's plates were circular, with a diameter of about the width of this page. They were separated by about 1.5 cm.

Before the Field is Switched on, there's only Gravity and the Viscous Force

1) With the electric field turned off, the forces acting on each oil drop are:

a) the **weight** of the drop — acting downwards
b) the **viscous force** from the air — acting upwards

Millikan had to take account of things like upthrust as well, but you don't have to worry about that — keep it simple.

2) The drop will reach **terminal velocity** (i.e. it will stop accelerating) when these two forces are equal. So, from Stoke's law (see above):

$$mg = 6\pi\eta rv$$

3) Since the **mass** of the drop is the **volume** of the drop multiplied by the **density**, ρ, of the oil, this can be rewritten as:

$$\frac{4}{3}\pi r^3 \rho g = 6\pi\eta rv \Rightarrow r^2 = \frac{9\eta v}{2\rho g}$$

Millikan measured η and ρ in separate experiments, so he could now calculate r — ready to be used when he switched on the electric field...

Millikan's Oil-Drop Experiment

Then he **Turned On** the **Electric Field**...

1) The field introduced a **third major factor** — an **electric force** on the drop.

2) Millikan adjusted the applied p.d. until the drop was **stationary**. Since the **viscous force** is proportional to the **velocity** of the object, once the drop stopped moving, the viscous force **disappeared**.

3) Now the only two forces acting on the oil drop were:

   a) the **weight** of the drop — acting downwards
   b) the force due to the **uniform electric field** — acting upwards

4) The **electric force** is given by: $F = \dfrac{QV}{d}$ where Q is the charge on the oil drop, V is the p.d. between the plates and d is the distance between the plates. **See p. 38–39**

5) Since the drop is **stationary**, this electric force must be equal to the weight, so:

$$\frac{QV}{d} = \frac{4}{3}\pi r^3 \rho g$$

The first part of the experiment gave a value for r, so the **only unknown** in this equation is Q.

6) So Millikan could find the **charge on the drop**, and repeated the experiment for hundreds of drops. The charge on any drop was always a **whole number multiple** of -1.6×10^{-19} C.

These Results Suggested that **Charge** was **Quantised**

1) This result was **really significant**. Millikan concluded that charge can **never exist** in **smaller** quantities than 1.6×10^{-19} C. He assumed that this was the **charge** carried by an **electron**.

2) Later experiments confirmed that **both** these things are true.

> Charge is "**quantised**". It exists in "packets" of size **1.6×10^{-19} C** — the **fundamental unit of charge**. This is the size of the charge carried by **one electron**.

Practice Questions

Q1 Write down the equation for Stoke's law, defining any variables.

Q2 List the forces that act on the oil drop in Millikan's experiment:
(a) with the drop drifting downwards at terminal velocity but with no applied electrical field,
(b) when the drop is stationary, with an electrical field applied.

Q3 Briefly explain the significance of Millikan's oil-drop experiment in the context of quantum physics.

Exam Questions

Q1 In Millikan's oil-drop experiment, an oil drop of mass 1.63×10^{-14} kg is stationary in the space between two horizontal plates 3.00 cm apart. The potential difference between the plates is 5000 V.

(a) Draw a free body force diagram for the drop. [2 marks]

(b) Neglecting the upthrust on the drop, calculate the size of the charge that it carries. Take $g = 9.81$ Nkg^{-1}. [3 marks]

(c) How many electrons did the oil drop lose or gain as it passed through the atomiser? [1 mark]

So next time you've got a yen for 1.59×10^{-19} coulombs — tough...

This was a huge leap. Along with the photoelectric effect (see p. 62) this experiment marked the beginning of quantum physics. The world wasn't ruled by smooth curves any more — charge now jumped from one allowed step to the next...

Light — Newton vs Huygens

These pages are for AQA A Option 8 only.

Newton was quite a bright chap really, but even he could make mistakes — and this was his biggest one. The trouble with being Isaac Newton is that everyone just assumes you're right...

Newton had his Corpuscular Theory

1) In 1672, Newton published his **theory of colour**. In it he suggested that **light** was made up of **tiny particles** that he called '**corpuscles**'.

2) One of his major arguments was that light was known to travel in **straight lines**, yet waves were known to **bend** in the shadow of an **obstacle** (diffraction). Experiments weren't **accurate enough** then to detect the diffraction of light. Light was known to **reflect** and **refract**, but that was it.

3) His theory was based on the principles of his **Laws of Motion** — that all particles, including his 'corpuscles', will 'naturally' travel in **straight lines**.

4) Newton believed that **reflection** was due to a force that **pushed** the particles away from the surface — just like a ball bouncing back off a wall.

5) **Refraction** worked if the corpuscles travelled **faster** in a **denser** medium.

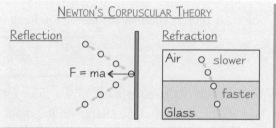

Huygens thought Light was a Wave

1) The idea that light might be a **wave** had existed for some time before it was formalised by Huygens in 1678.

2) At the time, nobody took much notice of him because his theory was **different** from Newton's.

3) Huygens developed a **general model** of the propagation of **waves** in what is now known as **Huygens' principle**. This states that:

> **HUYGENS' PRINCIPLE:** Every point on a wavefront may be considered to be a **point source** of **secondary wavelets** that spread out in the forward direction at the speed of the wave. The **new wave front** is the surface that is **tangential** to all of these **secondary wavelets**.

The diagram below shows how this works:

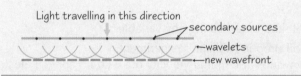

4) If he applied this theory to **light**, it could explain **reflection** and **refraction** easily.

Huygens predicted that light should **slow down** when it entered a **denser medium**, rather than speed up.

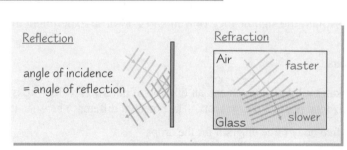

5) Huygens also predicted that light should **diffract** around tiny objects and that two coherent light sources should **interfere** with each other.

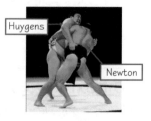

Up until the end of the 18th century, most scientists sided with **Newton**. He'd been right about so many things before, so it was generally assumed that he **must be right** about light being corpuscular. The debate raged for **over 100 years** until **Thomas Young** carried out experiments on the **interference** of light in Cambridge around **1800**...

Light — Newton vs Huygens

Young Proved Huygens Right with his Double-Slit Experiment

1) **Diffraction** and **interference** are both uniquely **wave** properties.

2) If it could be shown that **light** showed **interference** patterns, that would help decide once and for all between Newton's corpuscular theory and Huygens' wave theory.

3) The problem with this was getting two **coherent** (see p. 25) light sources. You **can't** arrange **two separate coherent light sources** because **light** from **each source** is emitted in **random bursts**.

4) Thomas Young solved this problem by using only **one point source of light** (in practice, a narrow slit with a filament lamp behind it). In front of the source, he put a **screen** with **two narrow slits** in it. Light spreading out by **diffraction** from the slits was equivalent to **two coherent point sources**.

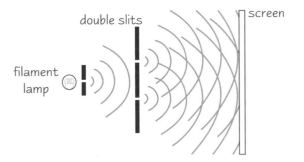

5) In the area on the screen where light from the two slits **overlapped**, bright and dark 'fringes' were formed.

6) This was **proof** that light could both **diffract** (through the narrow slits) and **interfere** (to form the interference pattern on the screen). **Huygens** was finally proved right... about 100 years after his death.

[See p. 30 for more details on the double-slit experiment.]

Observations and Theories Developed Rapidly during the 19th Century

1) In 1810, **polarisation** of light was seen for the first time. Physicists at the time thought that light spread like sound, as a longitudinal wave, so they struggled to explain polarisation..

2) In 1817, Young suggested that light was a **transverse wave**, explaining polarisation. Wave theory was now very widely accepted.

3) In the middle of the 19th century, James Clerk Maxwell showed theoretically that **all electromagnetic waves** should travel at the same speed in a vacuum, **c** (see page 22).

4) By that time, the **velocity of light** could be measured quite accurately, and it was found to be very close to Maxwell's value of **c**. This suggested that **light** is an **electromagnetic wave**.

5) This was the accepted theory up until the very end of the 19th century, when the **photoelectric effect** was discovered. Then the particle theory had to be resurrected, and it was all up in the air again...

Practice Questions

Q1   What was the main observation that Newton used to support his corpuscular theory of light?

Q2   What part does diffraction play in a Young's double-slit experiment?

Q3   Draw a diagram to show how you would set up an experiment to demonstrate Young's fringes for white light in a laboratory.

Exam Questions

Q1   Describe Newton's corpuscular theory of light.   [2 marks]

Q2   Give a brief history of the understanding of the nature of light in the 18th and 19th centuries. You should mention the various theories that have been proposed and the evidence that has been used to support them.   [5 marks]

In the blue corner — the reigning champion... IsaaaAAAAC NEWton...

So, light's a wave, right? We've got that well and truly sorted — the double-slit experiment laid the whole argument to rest once and for all. No arguments. There it is — light's a wave...   Or is it..? (Turn over for the next thrilling episode...)

The Photoelectric Effect

This page is for Edexcel and AQA B (AQA A Option 8 people — remind yourselves of this lot [you did it at AS]).

Shining Light on a Metal can Release Electrons

If you shine **light** of a **high enough frequency** onto the **surface of a metal**,
it will **emit electrons**. For **most** metals, this **frequency** falls in the **U.V.** range.

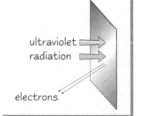

1) **Free electrons** on the **surface** of the metal **absorb energy** from the light, making them **vibrate**.

2) If an electron **absorbs enough** energy, the **bonds** holding it to the metal **break** and the electron is **released**.

3) This is called the **photoelectric effect** and the electrons emitted are called **photoelectrons**.

You don't need to know the details of any experiments on this — you just need to learn the three main conclusions:

Conclusion 1	For a given metal, **no photoelectrons are emitted** if the radiation has a frequency **below** a certain value — called the **threshold frequency**.
Conclusion 2	The photoelectrons are emitted with a variety of kinetic energies ranging from zero to some maximum value. This value of **maximum kinetic energy** increases with the **frequency** of the radiation, and is **unaffected** by the **intensity** of the radiation.
Conclusion 3	The **number** of photoelectrons emitted per second is **directly proportional** to the **intensity** of the radiation.

These are the two that had people puzzled. They can't be explained using wave theory.

The Photoelectric Effect Couldn't be Explained by Wave Theory

According to wave theory:

1) For a particular frequency of light, the **energy** carried is **proportional** to the **intensity** of the beam.

2) The energy carried by the light would be **spread evenly** over the wavefront.

3) **Each** free electron on the surface of the metal would gain a **bit of energy** from each incoming wave.

4) Gradually, each electron would gain **enough energy** to leave the metal.

SO... If the light had a **lower frequency** (i.e. was carrying less energy) it would take **longer** for the electrons to gain enough energy — but it would happen eventually. There is **no explanation** for the **threshold frequency**.

The **higher the intensity** of the wave, the **more energy** it should transfer to each electron — so the kinetic energy should increase with **intensity**. There's **no explanation** for the **kinetic energy** depending only on the **frequency**.

Einstein came up with the Photon Model of Light

1) When Max Planck was investigating **black body radiation** (don't worry, you don't need to know about that just yet), he suggested that **EM waves** can **only** be **released** in **discrete packets**, or **quanta**.

2) The **energy carried** by one of these **wave-packets** had to be:

$$E = hf = \frac{hc}{\lambda}$$

where h = Planck's constant = 6.63×10^{-34} Js
and c = speed of light in a vacuum = 3.00×10^8 ms^{-1}

3) **Einstein** went **further** by suggesting that **EM waves** (and the energy they carry) can only **exist** in discrete packets. He called these wave-packets **photons**.

4) He saw these photons of light as having a **one-on-one**, **particle-like** interaction with **an electron** in a **metal surface**. It would **transfer all** its **energy** to that **one, specific electron**.

The Photoelectric Effect

The **Photon Model** Explained the **Photoelectric Effect** Nicely

According to the photon model:
1) When light hits its surface, the metal is **bombarded** by photons.
2) If one of these photons **collides** with a free electron, the electron will gain energy equal to **hf**.

Before an electron can **leave** the surface of the metal, it needs enough energy to **break the bonds holding it there**. This energy is called the **work function energy** (symbol ϕ) and its **value** depends on the **metal**.

It Explains the **Threshold Frequency**...

1) If the energy **gained** from the photon is **greater** than the **work function energy**, the electron is **emitted**.
2) If it **isn't**, the electron will just **shake about a bit**, then release the energy as another photon. The metal will heat up, but **no electrons** will be emitted.
3) Since for **electrons** to be released, $hf \geq \phi$, the **threshold frequency** must be: $\boxed{f = \dfrac{\phi}{h}}$

... and the **Maximum Kinetic Energy**

1) The **energy transferred** to an electron is **hf**.
2) The **kinetic energy** it will be carrying when it **leaves** the metal is **hf minus** any energy it's **lost** on the way out (there are loads of ways it can do that, which explains the **range** of energies).
3) The **minimum** amount of energy it can lose is the **work function energy**, so the **maximum kinetic energy** is given by the equation: $\implies$ $\boxed{\dfrac{1}{2}mv_{max}^2 = hf - \phi}$
   The **kinetic energy** of the electrons is **independent of the intensity**, because they can **only absorb one photon** at a time.
4) The maximum kinetic energy can be **measured** using the idea of **stopping potential**. The emitted electrons are made to lose their **energy** by **doing work against** an applied **potential difference**. The work done by the p.d. in **stopping** the **fastest electrons** is equal to the energy they were carrying:

$$\boxed{\dfrac{1}{2}mv_{max}^2 = eV_s}$$

where **e** = charge on the electron = 1.6×10^{-19} C, V_s = stopping potential in V, and the maximum kinetic energy is measured in J. The kinetic energy in eV is just equal to the stopping potential — see p. 56.

Practice Questions

Q1 What three main conclusions were drawn from detailed experimentation on the photoelectric effect?
Q2 What is meant by the work function energy of a metal?
Q3 How is the energy of a photon related to its frequency?

Exam Questions *Use Planck's constant = 6.6×10^{-34} Js, speed of light = 3.0×10^8 ms^{-1}.*

Q1 An isolated zinc plate with neutral charge is exposed to high-frequency ultraviolet light. State and explain the effect of the ultraviolet light on the charge of the plate. [2 marks]

Q2 Light of wavelength 0.50 μm hits a metal surface and causes electrons to be released. Their maximum kinetic energy is 2.0×10^{-19} J.
 (a) Calculate the work function of the metal. [3 marks]
 (b) If a second beam of light causes no electrons to be emitted, find the lower limit of its wavelength. [2 marks]

Q3 Potassium has a work function of 2.2 eV. A potassium anode is illuminated with light of wavelength 350 nm. What potential must be applied to stop the electrons leaving the anode? [4 marks]

And that's all there is to it — *sob*...

Confused? The best way to really get your head round this sort of thing is to try and explain it to someone else. Learn why the light theory can't explain the photoelectric effect, and how photon theory does — then tell someone else about it.

Wave-Particle Duality

Theses pages are for Edexcel, AQA A Option 8 and AQA B only.

Is it a wave? Is it a particle?

Interference and Diffraction show Light as a Wave

1) Light produces **interference** and **diffraction** patterns — **alternating bands** of **dark** and **light**.

2) These can **only** be explained using **waves interfering constructively** (when two waves overlap in phase) or **interfering destructively** (when two waves are out of phase). (See p. 24.)

The Photoelectric Effect Shows Light Behaving as a Particle

1) **Einstein** explained the results of **photoelectricity experiments** (see p. 62) by thinking of the **beam of light** as a series of **particle-like "photons"**.

2) If a **photon** of light is a **discrete** bundle of energy, then it can **interact** with an **electron** in a **one-to-one way**.

3) **All** the **energy** in the **photon** is **given** to one **electron**.

Neither the **wave theory** nor the **particle theory** describe what light actually **is**. They're just two different **models** that help to explain the way light behaves.

I'm not impressed — this is just speculation. What do you think Dad?

De Broglie Came up With the Wave-Particle Duality Theory

1) Louis de Broglie made a **bold suggestion** in his **PhD thesis**:

> If **"wave-like" light** showed **particle properties** (photons), **"particles"** like **electrons** should be expected to show **wave-like properties**.

2) The **de Broglie equation** relates a **wave property** (**wavelength, λ**) to a **moving particle property** (**momentum, mv**). **h** = Planck's constant = 6.63 × 10⁻³⁴ Js.

$$\lambda = \frac{h}{mv}$$

Most physicists at the time weren't very impressed — his ideas were just speculation.

3) The **de Broglie wave** of a particle can be interpreted as a **"probability wave"**. The **probability** of finding a particle at a point is **directly proportional** to (*amplitude of the de Broglie wave*)².

4) Later experiments **confirmed** the wave nature of electrons.

Electron Diffraction shows the Wave Nature of Electrons

1) De Broglie's suggestions prompted a lot of experiments to try to show that **electrons** can have **wave-like** properties. In **1927**, Davisson and Germer succeeded in **diffracting electrons**.

2) They saw **diffraction patterns** when **accelerated electrons** in a vacuum tube **interacted** with the **spaces** in a graphite **crystal**.

3) According to wave theory, the **spread** of the **lines** in the diffraction pattern **increases** if the **wavelength** of the wave is **greater**.

4) In electron diffraction experiments, a **smaller accelerating voltage**, i.e. **slower** electrons, gives **widely spaced** rings.

5) **Increase** the **electron speed** and the diffraction pattern circles **squash together** towards the **middle**. This fits in with the **de Broglie** equation above — if the **velocity** is **higher**, the **wavelength** is **shorter** and the **spread** of the lines is **smaller**.

Electron diffraction patterns look like this.

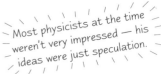

For AQA A astrophysics students, this circle is called the Airy disc (see p. 111).

In general, λ for **electrons** accelerated in a **vacuum tube** is about the **same size** as **electromagnetic waves** in the **X-ray** part of the spectrum.

6) **Just for AQA A Option 8**, the de Broglie wavelength of an electron (**λ**) is related to the **accelerating voltage** (**V**) by:

$$\lambda = \frac{h}{\sqrt{2meV}}$$

where **e** is the charge on the electron and **m** is its mass

Wave-Particle Duality

Particles Don't Show Wave-Like Properties All the Time

You **only** get **diffraction** if a particle interacts with an object of about the **same size** as its **de Broglie wavelength**.
A **tennis ball**, for example, with **mass 0.058 kg** and **speed 100 ms⁻¹** has a **de Broglie wavelength** of 10^{-34} m.
That's 10^{19} **times smaller** than the **nucleus** of an **atom**! There's nothing that small for it to interact with.

> ### Example
> An electron of mass 9×10^{-31} kg is fired from an electron gun at 7×10^6 ms⁻¹.
> What size object will the electron need to interact with in order to diffract?
>
> Momentum of electron = mv = 6.3×10^{-24} kg ms⁻¹
> $\lambda = h/mv = 6.63 \times 10^{-34} / 6.3 \times 10^{-24} = \boxed{1 \times 10^{-10} \text{ m}}$ → Only crystals with atom layer spacing around this size are likely to cause the diffraction of this electron.

A **shorter wavelength** gives **less diffraction**. This is important in **microscopes** where diffraction **blurs out details**.
The **tiny** wavelength of electrons means an **electron microscope** can resolve **finer detail** than a **light** microscope.

Electron Microscopes use Electrons Instead of Light   *AQA A Option 8 only*

In electron microscopes:

1) A **stream of electrons** is accelerated towards the sample using a **positive electric potential**.

2) This stream is confined into a thin **beam** using a **magnetic field**.

3) The beam is **focused** onto the sample and any interactions are transformed into an **image**.
The sort of image you get depends on the **type of microscope** you're using:

> A **transmission electron microscope** (**TEM**) works a bit like a **slide** projector, but uses electrons instead of light. A **very thin** specimen is used and the parts of the beam that pass through the specimen are projected onto a **screen** to form an image.

> A **scanning tunnelling microscope** (**STM**) uses principles of **quantum mechanics**. A very fine **probe** is moved over the surface of the sample and a **voltage** is applied between the probe and the surface. Electrons "tunnel" from the probe to the surface, resulting in a weak **electrical current**. The smaller the **distance** between the probe and the surface, the **greater the current**. By scanning the probe over the surface and measuring the current, you produce an **image** of the **surface** of the sample.

Practice Questions

Q1   What name is normally given to "particles" of light?
Q2   What observation showed that electrons could behave as waves?
Q3   What is the advantage of an electron microscope over a light microscope?

Exam Questions

Use $h = 6.63 \times 10^{-34}$ Js ; $c = 3.00 \times 10^8$ ms⁻¹.

Q1   An electron is accelerated through a p.d. of 500 V.
  (a) Given that its mass is 9.1×10^{-31} kg and its charge is -1.6×10^{-19} C, calculate:
    i) the velocity of the electron,   ii) its de Broglie wavelength.   [4 marks]
  (b) In which region of the electromagnetic spectrum does this fall?   [1 mark]

Q2   Find an expression for the de Broglie wavelength of a particle in terms of its
    kinetic energy, E_K, its mass, m, and Planck's constant, h.   [3 marks]

Wave-Particle duelity — pistols at dawn...

*Anyone doing AQA A will have seen a lot of this before at AS, but there are quite a few extra details this time round.
You're getting into the weird bits of quantum physics now — it says that light isn't a wave, and it isn't a particle, it's **both**...
at the **same time**. It's what quantum physicists like to call a "juxtaposition of states" — well, they would, wouldn't they...*

Electron Energy Levels

These pages are for Edexcel, AQA B and OCR B only.

Electrons in Atoms Exist in Discrete Energy Levels

1) **Electrons** in an **atom** can **only exist** in certain **well-defined energy levels**. Each level is given a **number** (called the **principal quantum number** of the electron in that state), with $n = 1$ representing the electron's lowest possible energy — its **ground state**.

2) Electrons can **move down** an energy level by **emitting** a **photon**.

3) Since these **transitions** are between **definite energy levels**, the **energy** of **each photon** emitted can **only** take **certain values**.

4) The diagram on the right shows the **energy levels** for **atomic hydrogen**.

5) The **energies involved** are **so tiny** that it makes sense to use a more **appropriate unit** than the **joule**. The **electron-volt (eV)** (defined on page 56) is used instead. On the diagram, energies are labelled in **both units** for **comparison**.

6) The **energy** carried by each **photon** is **equal** to the **difference in energies** between the **two levels**. The equation below shows a **transition** between levels $n = 2$ and $n = 1$:

$$\Delta E = E_2 - E_1 = hf = \frac{hc}{\lambda}$$

7) In the same way, atoms can only **absorb** allowed photon-energies. This **quantisation** of electron energies in atoms produces **line emission** and **absorption spectra** (see p. 114).

LEVEL	ENERGY
$n = \infty$	zero energy
$n = 5$	-8.6×10^{-20} J or -0.54 eV
$n = 4$	-1.4×10^{-19} J or -0.85 eV
$n = 3$	-2.4×10^{-19} J or -1.5 eV
$n = 2$	-5.4×10^{-19} J or -3.4 eV
$n = 1$	-2.2×10^{-18} J or -13.6 eV

transitions

The energies are only negative because of how "zero energy" is defined. Just one of those silly convention things — don't worry about it.

> Electrons (like protons and neutrons) are examples of **fermions**. That means they obey the **Pauli exclusion principle**. This states that **no two fermions** can be in **exactly** the same **quantum state** at the same time. In the context of energy levels, that means **no more than two** electrons can be in the same **energy level** at the same time. (*Two are allowed at each energy — one spin-up, one spin-down. You don't need to know about spin though.*)

Lasers work by Stimulated Photon Emission

AQA B only

1) An important **application** of this stuff is the **laser** (**l**ight **a**mplification by **s**timulated **e**mission of **r**adiation).

2) Normally, electrons only stay in **excited states** for **very short periods** of time before dropping back to a lower level and emitting a photon (called **spontaneous emission**).

3) **Stimulated emission** happens when a photon of exactly the right frequency passes an atom with an electron in an excited state. This causes the electron to emit a photon of the **same frequency** and **in phase** with the first.

4) These two can go on to stimulate **further** emission from other excited atoms, producing a **burst** of photons that are all **in phase** and at the **same frequency**.

Stimulated Emission

hf

$E + hf$

E

hf
hf

5) Materials that can be used for **lasers** have a **metastable** (stable-ish) **state** above the ground state where electrons can get "**trapped**" for relatively long times — about a millisecond or so.

6) In a laser, electrons are "**pumped**" up to **very high energies** (exactly how that's done depends on the type of laser). These high-energy electrons decay **spontaneously** down to the **metastable state** until there are more electrons in the metastable state than the ground state — a **population inversion**.

7) Some of these electrons will then fall to the **ground state** spontaneously and **stimulate** other excited atoms to emit photons **in phase** with them.

8) By putting the material in which this is happening between **two mirrors**, photons are made to **bounce** back and forth, producing **further stimulated emission** and **amplification**.

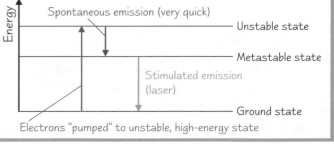

Energy

Spontaneous emission (very quick) ⟶ Unstable state

Metastable state

Stimulated emission (laser)

Ground state

Electrons "pumped" to unstable, high-energy state

Electron Energy Levels

Lasers have Loads of Uses — *AQA B only*

1) <u>Bar codes</u> — **reflected laser light** is used to detect the width of black lines on bar codes.

2) <u>Communication</u> — semiconductor lasers produce light in the **near infrared**, which isn't absorbed much by glass. This means information can be transmitted over **very large distances** using optical fibre.

3) <u>Medical and industrial uses</u> — the energy of the laser beam is **absorbed** by the target material in a very localised area, making **microsurgery** and **accurate cutting** possible.

4) <u>Laser printers</u> — a laser produces an **electrostatic image** on a plate or drum. Toner powder sticks to the areas that are charged and can be transferred by contact onto paper, then fixed by heating.

5) <u>Information storage and retrieval</u> — CDs and DVDs. Information is stored in **digital form** on tracks that spiral out from the centre of the disc and contain a series of **pits**. Laser light is focused on the tracks and is reflected from the top and bottom of the pits, which are **¼ wavelength** deep. Destructive interference occurs, and the reflected light is interpreted as a digital electronic signal.

The Wave Model of the Atom can Help you Understand Energy Levels

1) Since light has both **particle** and **wave** characteristics (see p. 64), de Broglie suggested that **electrons** should have a **wave-like character**.

2) Specifically, when they're in orbit **around a nucleus** they ought to behave like the **standing waves** that are formed on a guitar string when it's plucked (see p. 26).

3) Just as standing waves on the guitar string only exist at certain **well-defined frequencies**, only certain standing waves are possible in an atom.

4) For the **circular orbits** suggested by **Bohr**, the **wavelength** of the electron waves should fit the **circumference** a **whole number** of times.

5) The **principal quantum number** (corresponding to the number of the energy level) is equal to the number of **complete waves** that fit the circumference.

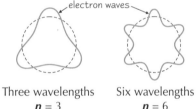

electron waves

Three wavelengths
n = 3

Six wavelengths
n = 6

Not a standing wave
Forbidden energy

Bohr wasn't actually thinking along these lines when he came up with his model of the atom — he just found some rules that worked. This interpretation came along much later.

Practice Questions

Q1 What is the difference between spontaneous emission and stimulated emission? *(AQA B only)*

Q2 List three common uses of lasers. *(AQA B only)*

Q3 Describe the standing-wave model of electrons in an atom.

Q4 State Pauli's exclusion principle. What type of particle does it apply to?

Exam Questions

Q1 The second is defined using the radiation from a particular quantum jump in the caesium-133 atom. The difference in energy levels is 3.8×10^{-5} eV.

(a) Calculate the frequency of this radiation ($h = 6.6 \times 10^{-34}$ Js). [2 marks]

(b) How many oscillations occur in 1 second, as used in the definition of the second? [1 mark]

Q2 Explain briefly the principles behind the operation of a laser. [5 marks]

Baked beans can cause stimulated emissions...

Soooo... that was quantum physics. It wasn't so bad was it? Ahem.

Ideal Gases

This page is for AQA B, OCR A and OCR B only.

*Aaahh... great... another one of those 'our equation doesn't work properly with **real gases**, so we'll invent an **ideal** gas that it **does work** for and they'll think we're dead clever' situations. Hmm. Physicists, eh...*

ALL equations in **thermal physics** use temperatures measured in kelvin.

The Kelvin (Thermodynamic) Temperature Scale

1)   A change of **1 K** equals a change of **1 °C**.
2)   To change from degrees Celsius into kelvin **add 273.15**: $K = C + 273.15$

There's more on where this scale comes from on page 70. You'll need to use kelvin for the whole of this section.

There are **Three Gas Laws**

The three gas laws were each worked out **independently** in the 17th century, by **careful experiment**. Each of the gas laws applies to a **fixed mass** of gas.

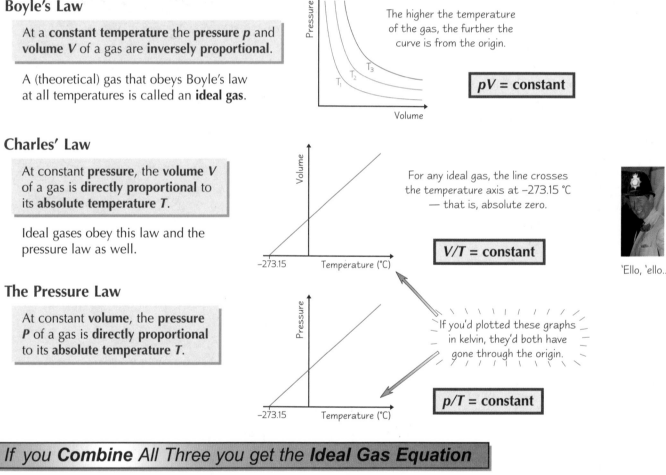

Boyle's Law

At a **constant temperature** the pressure **p** and volume **V** of a gas are **inversely proportional**.

A (theoretical) gas that obeys Boyle's law at all temperatures is called an **ideal gas**.

The higher the temperature of the gas, the further the curve is from the origin.

$pV = \text{constant}$

Charles' Law

At constant **pressure**, the **volume V** of a gas is **directly proportional** to its **absolute temperature T**.

Ideal gases obey this law and the pressure law as well.

For any ideal gas, the line crosses the temperature axis at –273.15 °C — that is, absolute zero.

$V/T = \text{constant}$

'Ello, 'ello...

The Pressure Law

At constant **volume**, the **pressure P** of a gas is **directly proportional** to its **absolute temperature T**.

If you'd plotted these graphs in kelvin, they'd both have gone through the origin.

$p/T = \text{constant}$

If you **Combine** All Three you get the **Ideal Gas Equation**

Combining all three gas laws gives the equation: $\dfrac{pV}{T} = \text{constant}$

1)   The constant in the equation depends on the amount of gas used.   *(Pretty obvious... if you have more gas it takes up more space.)*
     The amount of **gas** can be **measured** in moles, **n**.

2)   The seconstant then becomes **nR**, where **R** is called the **molar gas constant**.
     Its value is 8.31 J mol⁻¹ K⁻¹.

3)   Plugging this into the equation gives: $\dfrac{pV}{T} = nR$ or rearranging, $pV = nRT$ — *the ideal gas equation*

This equation works well (i.e., a real gas approximates to an ideal gas) for gases at **low pressure** and fairly **high temperatures**.

Ideal Gases

This page is for AQA B and OCR B only.

An *Ideal Gas* Obeys the *Assumptions* of *Kinetic Theory*

Kinetic theory tries to explain the gas laws by treating a gas as a series of **hard balls** moving at **high speed** in **random directions**. You can get loads of useful equations from the theory (p. 70-71), if you use some **simplifying assumptions**:

1) The gas contains a **large number of particles**.
2) The particles **move rapidly** and **randomly**.
3) The motion of the particles follows **Newton's laws**.
4) **Collisions** between particles themselves or at the walls of a container are **perfectly elastic**.
5) There are **no attractive forces** between particles.
6) Any **forces** that act during collisions are **instantaneous**.
7) Particles have a **negligible volume** compared with the volume of the container.

Any **gas obeying** these **assumptions** is an **ideal** gas. Real gases behave like ideal gases as long as the **pressure isn't too big** and the **temperature** is **reasonably high** (compared with the boiling point of the gas).

Brownian Motion Supports Kinetic Theory

In 1827, the botanist **Robert Brown** was looking at pollen grains in water. He noticed that they constantly moved with a zigzag, **random motion**.

Brownian Motion Experiment

You can **observe** Brownian motion in the lab.

Put some **smoke** in a **brightly illuminated** glass jar and observe the particles using a **microscope**.

The smoke particles appear as **bright specks** moving **haphazardly** from side to side, and up and down.

Brown couldn't explain this, but nearly 80 years later Einstein used it as evidence for the existence of **molecules** in the air. The **randomly moving** molecules were hitting the smoke particles unevenly, causing the motion.

Practice Questions

Q1   The pressure of a gas is 100 000 Pa and its temperature is 27 °C. The gas is heated — its volume stays fixed but the pressure rises to 150 000 Pa. Show that its new temperature is 177 °C.

Q2   What seven assumptions are made about ideal gas behaviour in kinetic theory?

Exam Questions

Q1   The mass of one mole of nitrogen gas is 0.028 kg. R = 8.31 J mol⁻¹ K⁻¹.

(a)   A flask contains 0.014 kg of nitrogen gas. How many moles of nitrogen are in the flask?   [1 mark]

(b)   If the flask is of volume 0.01 m³ and is at a temperature of 27 °C, what is the pressure inside it?   [2 marks]

Q2   A large helium balloon has a volume of 10 m³ at ground level. The temperature of the gas in the balloon is 293 K and the pressure is 1 × 10⁵ Pa. The balloon is released and rises to a height where its volume becomes 25 m³ and its temperature is 260 K.

(a)   Calculate the pressure inside the balloon at its new height.   [3 marks]

(b)   State two assumptions you made in your calculations.   [2 marks]

Brownian motion — girl guide in a tumble-dryer...

All this might sound a bit theoretical, but most gases you'll meet in the everyday world come fairly close to being 'ideal'. They only stop obeying these laws when the pressure's too high or they're getting close to their boiling point.

Molecular Kinetic Theory & Internal Energy

These pages are for AQA B, OCR A and OCR B only.

*The energy of a particle depends on its temperature on the **thermodynamic scale** (that's kelvin to you and me).*

The **Thermodynamic** Temperature Scale uses the **Triple Point** of Water

The thermodynamic scale is defined in terms of **two fixed points**: **absolute zero** and the **triple point of water**.

1) **Absolute zero** is the **lowest possible temperature**, and something at this temperature has the **lowest possible internal energy**. This is **zero kelvin**, written **0 K**, on the thermodynamic (absolute) temperature scale.

2) The **triple point of water** (subscript *tr* in the equation) is the temperature when ice, water and water vapour are in **thermal equilibrium** (they all exist at the same time). This is given a value of **273.16 K** on the absolute scale — so that a change of 1 K is equal to a change of 1 °C.

> *Just for AQA B*, the temperature, *T*, in kelvin of a substance at pressure *p* and with volume *V* is: $T = \dfrac{pV}{(pV)_{tr}} \times 273.16$

The **Speed** of a Particle Depends on its **Temperature**

The **particles** in a gas **don't** all **travel** at the **same speed**. At any given time, most will travel at around the **average speed**, but some will move much faster or slower. The shape of the **speed distribution** depends on the **temperature** of the gas.

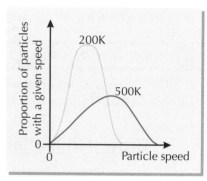

> As the temperature of the gas increases:
> 1) the **average** particle speed increases.
> 2) the **maximum** particle speed increases.
> 3) the distribution curve **spreads out**.

The particles of a gas **collide** with each other **all the time**. Some of these are '**head-on**' (particles moving in **opposite directions**) and others are '**shunts from behind**' (particles moving in the **same direction**).

1) As a result of the collisions, **energy's transferred** from one particle to another.

2) Some particles **gain speed** in a collision and others **slow down**.

3) Even though the energy of an individual particle changes at each collision, the collisions **don't alter** the **total energy** of the **system**.

4) So, the **average speed** of the particles stays the same so long as the **temperature** of the gas **stays the same**.

The **Average Kinetic Energy** of a Particle is **Proportional to T**

There are **two equations** for the **product *pV*** of a gas — the ideal gas equation, and an equation derived from kinetic theory (you don't need to know the derivation). You can **equate these** to get an expression for the **average kinetic energy**.

1) The **ideal gas equation**: $pV = nRT$

2) The **pressure** of an **ideal gas** given by kinetic theory: $pV = \dfrac{1}{3}Nm\overline{c^2}$

> *c* is the velocity of a particle. $\overline{c^2}$ is the average of the squared speeds of the particles, called the <u>mean square speed</u>.
>
> *N* is the number of <u>molecules</u> in the gas.
>
> *m* is the mass of <u>one molecule</u> in the gas.

3) **Equating** these two gives: $\dfrac{1}{3}Nm\overline{c^2} = nRT$

4) **Multiplying** by 3/2 gives: $\dfrac{3}{2} \times \dfrac{1}{3}Nm\overline{c^2} = \dfrac{3nRT}{2}$

so: $$\dfrac{1}{2}m\overline{c^2} = \dfrac{3}{2}\dfrac{nRT}{N}$$

5) $\dfrac{1}{2}m\overline{c^2}$ is the **average kinetic energy** of an **individual particle**.

So the average kinetic energy of the gas particles is **directly proportional** to the **absolute temperature**, *T*.

6) The **internal energy** of the gas is the sum of all the **potential** and **kinetic** energies of the particles.

Molecular Kinetic Theory & Internal Energy

Boltzmann's Constant k is Equal to R/N_A

One mole of any **gas** contains the same number of particles. This number is called **Avogadro's constant** and has the symbol N_A. The value of N_A is **6.02 × 10²³ particles per mole**.

1) The **number of particles** in a **mass of gas** is given by the **number of moles**, *n*, multiplied by **Avogadro's constant**.
   So the number of particles, $N = nN_A$

2) This can be used to **simplify** the **kinetic energy equation** from page 70 to:
   $$\frac{1}{2}m\overline{c^2} = \frac{3RT}{2N_A}$$
   where *m* = mass of one molecule

3) **Boltzmann's constant**, *k* is equivalent to R/N_A, so:
   $$\frac{1}{2}m\overline{c^2} = \frac{3}{2}kT$$

4) The value of Boltzmann's constant is **1.38 × 10⁻²³ JK⁻¹**.

5) You can think of Boltzmann's constant as the **gas constant** for **one particle of gas**, while **R** is the gas constant for **one mole of gas**.

A Useful Quantity in Kinetic Theory is the Root Mean Square Speed √c̄²

In kinetic theory, it helps to think about the motion of a typical particle.
1) $\overline{c^2}$ is the **average** of the **squared speeds** of the particles, so the square root of it gives you the **typical** speed.
2) This is called the **root mean square speed** or, usually, the **r.m.s. speed**.

$$r.m.s.\ speed = \sqrt{mean\ square\ speed} = \sqrt{\overline{c^2}}$$

Note — not the average speed.

Practice Questions

Q1 Describe the changes in the distribution of gas particle speeds as the temperature of a gas increases.

Q2 Explain how the speed of a gas particle can change even though the temperature of the gas stays constant.

Q3 What happens to the average kinetic energy of a particle if the temperature of a gas doubles?

Exam Questions

Q1 The mass of one mole of nitrogen molecules is 2.8 × 10⁻² kg. There are 6.02 × 10²³ molecules in one mole.

(a) What is the mass of one molecule? [1 mark]

(b) Calculate the typical speed of a nitrogen molecule at 300 K. [3 marks]

Q2 Some air freshener is sprayed at one end of a room. The room is 8.0 m long and the temperature is 20 °C.

(a) Assuming the average freshener molecule moves at 400 ms⁻¹, how long will it take for a particle to travel directly to the other end of the room? [1 mark]

(b) The perfume from the air freshener only slowly diffuses from one end of the room to the other. Explain why this takes much longer than suggested by your answer to part (a). [2 marks]

(c) What difference would you notice about the speed of diffusion for the air freshener if the temperature was 30 °C? [2 marks]

Positivise your internal energy, man...

Phew... there's a lot to take in on these pages. Go back over it, step by step, and make sure you understand it all: the absolute temperature scale, average kinetic energy of a gas, Boltzmann's constant and the r.m.s. speed.

Specific Heat Capacity & Specific Latent Heat

This page is for AQA B, OCR A and OCR B only (OCR B people don't need to worry about the experiments).

You need energy to heat something up, and to change its state. Everything comes down to energy. Pretty much always.

Specific Heat Capacity *is how much* Energy *it Takes to* Heat *Something*

When you heat something, its molecules get more **kinetic energy** and its **temperature** rises.

> The **specific heat capacity** (**c**) of a substance is the amount of **energy**
> needed to **raise** the **temperature** of **1 kg** of the substance by **1 K** (or 1°C).

or put another way: | **energy change = mass × specific heat capacity × change in temperature**

in symbols: | $\Delta Q = mc\Delta\theta$ ⟵ ΔT is sometimes used instead of $\Delta\theta$ for the change in temperature.

ΔQ is the energy change in J, **m** is the mass in kg and $\Delta\theta$ is the temperature change in K or °C. Units of **c** are $J\,kg^{-1}\,K^{-1}$ or $J\,kg^{-1}\,°C^{-1}$.

You can Measure *Specific Heat Capacity in the* Laboratory

The **method**'s the same for **solids** and **liquids**, but the **set-up**'s a little bit different (you don't have to memorise these experiments):

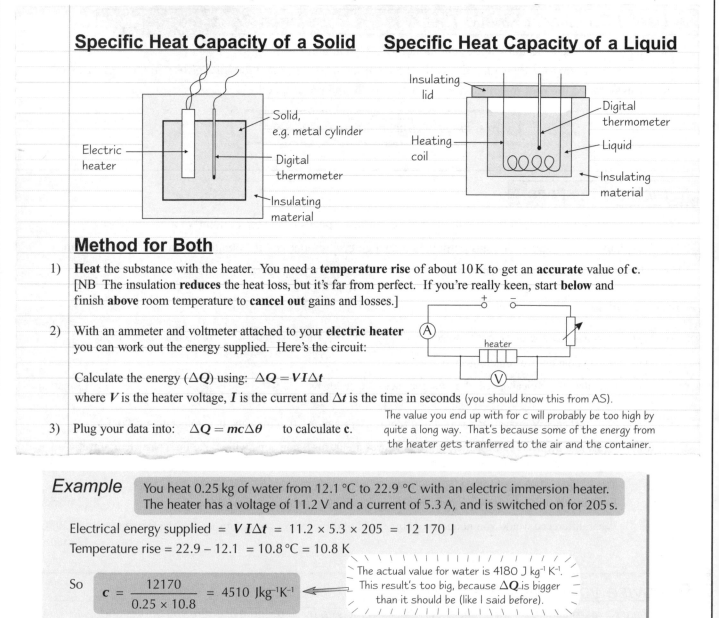

Specific Heat Capacity of a Solid

Electric heater — Solid, e.g. metal cylinder
Digital thermometer
Insulating material

Specific Heat Capacity of a Liquid

Insulating lid
Heating coil
Digital thermometer
Liquid
Insulating material

Method for Both

1) **Heat** the substance with the heater. You need a **temperature rise** of about 10 K to get an **accurate** value of **c**.
[NB The insulation **reduces** the heat loss, but it's far from perfect. If you're really keen, start **below** and finish **above** room temperature to **cancel out** gains and losses.]

2) With an ammeter and voltmeter attached to your **electric heater** you can work out the energy supplied. Here's the circuit:

Calculate the energy (ΔQ) using: $\Delta Q = VI\Delta t$

where V is the heater voltage, I is the current and Δt is the time in seconds (you should know this from AS).

3) Plug your data into: $\Delta Q = mc\Delta\theta$ to calculate **c**.

The value you end up with for c will probably be too high by quite a long way. That's because some of the energy from the heater gets transferred to the air and the container.

Example

> You heat 0.25 kg of water from 12.1 °C to 22.9 °C with an electric immersion heater.
> The heater has a voltage of 11.2 V and a current of 5.3 A, and is switched on for 205 s.

Electrical energy supplied = $VI\Delta t$ = 11.2 × 5.3 × 205 = 12 170 J

Temperature rise = 22.9 – 12.1 = 10.8 °C = 10.8 K

So | $c = \dfrac{12170}{0.25 \times 10.8} = 4510\ Jkg^{-1}K^{-1}$ ⟵

The actual value for water is 4180 $J\,kg^{-1}\,K^{-1}$. This result's too big, because ΔQ is bigger than it should be (like I said before).

Specific Heat Capacity & Specific Latent Heat

It takes *Energy* to *Change State*

To **melt** a **solid**, you need to **break the bonds** that hold the molecules in place. The **energy** needed for this is called the **latent heat of fusion**. Similarly, when you **evaporate a liquid**, you have to **put in energy** to **pull the molecules apart** completely. This is the **latent heat of vaporisation**.

Specific Latent Heat is Defined as the Latent Heat *per kg*

The **larger** the **mass** of the substance, the **more energy** it takes to **change** its **state**. That's why the **specific latent heat** is defined per kg:

> The **specific latent heat (l)** of **fusion** or **vaporisation** is the quantity of **thermal energy** required to **change the state** of **1 kg** of a substance.

which gives:

> **energy change = specific latent heat × mass of substance changed**

or in symbols:

$$\Delta Q = l\Delta m$$

You'll usually see the latent heat of vaporisation written l_v and the latent heat of fusion written l_f.

Where ΔQ is the energy change in J and Δm is the mass in kg. The units of l are J kg^{-1}.

Measuring *Specific Latent Heat* is Similar to Measuring *c*

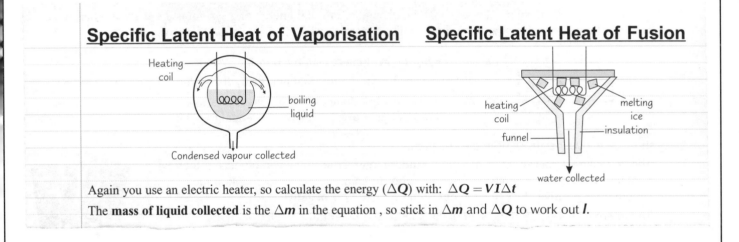

Specific Latent Heat of Vaporisation

Heating coil
boiling liquid
Condensed vapour collected

Specific Latent Heat of Fusion

heating coil
melting ice
funnel
insulation
water collected

Again you use an electric heater, so calculate the energy (ΔQ) with: $\Delta Q = VI\Delta t$

The **mass of liquid collected** is the Δm in the equation, so stick in Δm and ΔQ to work out l.

Practice Questions

Q1 Define specific heat capacity.

Q2 Describe how you would measure the specific heat capacity of olive oil.

Q3 Show that the thermal energy needed to heat 2 kg of water from 20 °C to 50 °C is ~250 kJ (c_{water} = 4180 Jkg^{-1}K^{-1}).

Q4 Explain why energy is needed to evaporate a liquid and define the specific latent heat of vaporisation.

Exam Questions

Q1 In an experiment to measure the specific heat capacity of a metal, a 2 kg metal cylinder is heated uniformly from 4.5 °C to 12.7 °C in 3 minutes. The electric heater used is rated at 12 V, 7.5 A. Assuming that heat losses were negligible, calculate the specific heat capacity of the metal. [3 marks]

Q2 A 3 kW electric kettle contains 0.5 kg of water already at its boiling point. Neglecting heat losses, how long will it take to boil dry? (Specific latent heat of vaporisation of water = 2.26 × 10^6 J kg^{-1}) [3 marks]

My specific eat capacity — 24 pies...

*This stuff's a bit dull, but hey... make sure you're comfortable using those equations. Interesting(ish) fact for the day — it's the **huge** difference in specific heat capacity between the land and the sea that causes the monsoon in Asia. So there.*

Thermodynamics

These pages are for AQA B only

The 1ˢᵗ and 2ⁿᵈ laws had been around for ages before someone realised they'd missed one — hence the 0ᵗʰ law. Doh!

The **Zeroth Law** of **Thermodynamics** is about **Thermal Equilibrium**

The **zeroth law** of thermodynamics says that if **body A** and **body B** are both in **thermal equilibrium** with **body C**, then **body A** and **body B** must be in thermal equilibrium with **each other**.

This is linked with the idea of **temperature**.

1) Suppose A, B and C are three identical metal blocks. A has been in a **warm oven**, B has come from a **refrigerator** and C is at **room temperature**.

2) **Thermal energy** flows from A to C and C to B until they all reach **thermal equilibrium** and the net flow of energy stops. This happens when the three blocks are at the **same temperature**.

The **First Law** of **Thermodynamics** is **Q = ΔU – W**

There are **two** ways to raise the **temperature** of a system:

1) Put it in **thermal contact** with something **hotter** (i.e. supply heat energy)
2) **Do work** on it. E.g. you can raise the temperature of your arm by rubbing it, i.e. doing work against friction.

Temperature is a measure of how much **internal energy** a system has.

The **first law** of thermodynamics says that the **change in internal energy** of a **system** is the **thermal energy put into the system** minus the **work done by the system** — it's just conservation of energy.

$$Q = \Delta U - W$$

1) Q is the **heat energy** supplied **to** the system.
2) ΔU is the **gain in internal energy** by the system.
3) W is the **work done on** the system. For a **constant pressure**, p, this work is equal to $p\Delta V$.

Make sure you get your **signs right** in the equation:

1) The flow of **heat energy**, Q, comes from **random interactions** between **objects** in **thermal contact**. The heat energy **always flows** from **hot to cold**. If heat is flowing **into** the system, Q **is positive**, if it's flowing **out** of the system, Q **is negative**.

2) The **internal energy**, U, is the **sum** of the **kinetic** and **potential** energies of all the particles. If the internal energy **rises**, ΔU **is positive**, and if it **falls**, ΔU **is negative**.

3) The **work done** is due to any process that's **independent** of the **temperature difference**. If work is done **on** the system, W **is positive**, if work is done **by** the system, W **is negative**.

Getting your signs right is generally a good idea.

Some **Energy Changes** are **Adiabatic**, **Isothermal** or **Constant Volume**

You need to know about **three special cases** of energy changes when using the **first law** of thermodynamics.

Adiabatic change — no heat goes in, no heat comes out
When you put $Q = 0$ into the first law equation, you get $0 = \Delta U - W \Rightarrow \Delta U = W$
So if **work** is done **on** the system, the **internal energy rises** and the **temperature** goes **up**.
If **work** is done **by** the system, the **internal energy falls** and the **temperature** goes **down**.

Isothermal change — the temperature stays the same
This time, $\Delta U = 0$, so $Q = -W$
So the **heat** going **into** the system equals the **work** being done **by** the system.

Constant volume — no work is done (since $\Delta V = 0$ and $W = p\Delta V$)
$W = 0$, so the first law becomes $Q = \Delta U$
So if energy goes **into** the system, the **internal energy increases** and the temperature **rises**. And vice versa.

Thermodynamics

Compression and Expansion of Gases

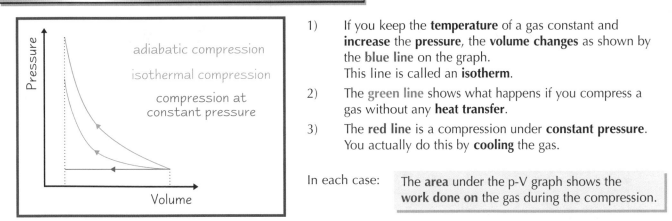

1) If you keep the **temperature** of a gas constant and **increase** the **pressure**, the **volume changes** as shown by the **blue line** on the graph.
This line is called an **isotherm**.

2) The **green line** shows what happens if you compress a gas without any **heat transfer**.

3) The **red line** is a compression under **constant pressure**.
You actually do this by **cooling** the gas.

In each case:

The **area** under the p-V graph shows the **work done on** the gas during the compression.

If the gas is then allowed to **expand** under the **same conditions**, the expansion will follow the **same path** in reverse.

Heat Engines use Expansion and Compression Cycles

Heat engines absorb heat from a source, produce some **mechanical work** and **lose** some **heat** energy to an **exhaust**. The graph shows how the pressure and volume of a gas change in an **ideal heat engine** called the **Carnot cycle**.

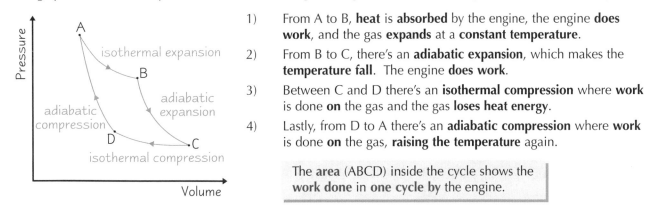

1) From A to B, **heat is absorbed** by the engine, the engine **does work**, and the gas **expands** at a **constant temperature**.

2) From B to C, there's an **adiabatic expansion**, which makes the **temperature fall**. The engine **does work**.

3) Between C and D there's an **isothermal compression** where **work** is done **on** the gas and the gas **loses heat energy**.

4) Lastly, from D to A there's an **adiabatic compression** where **work** is done **on** the gas, **raising the temperature** again.

The **area** (ABCD) inside the cycle shows the **work done** in **one cycle by** the engine.

Practice Questions

Q1   State the zeroth law of thermodynamics.

Q2   Write down two ways to increase the temperature of a system.

Q3   What is meant by i) an isothermal change, ii) an adiabatic change?

Q4   A gas is compressed and a graph plotted of volume vs. pressure. What does the area under this graph represent?

Exam Questions

Q1 (a) Write down an equation to represent the first law of thermodynamics, carefully defining each
term in the equation.   [3 marks]

(b) If 60 J of work is done on a system but the system loses 20 J of energy in heat,
what will be the change in the internal energy of the system?   [2 marks]

(c) Will the temperature rise, fall or remain constant during this change?   [1 mark]

Adiabatic — no heat transfer...   Isothermal — constant temperature...

Okay... the 0th law's a bit obvious, really. It just says that any objects that are in thermal equilibrium are at the same temperature. For the 1st law, make sure you learn the special cases — adiabatic, isothermal and constant volume.

The Boltzmann Factor

These pages are for OCR B only.

Welcome to the big bad world of statistical physics — Ludwig Boltzmann's got a lot to answer for...

The Thermal **Energy** of a Particle is Proportional to the **Temperature**

1) Any particle above absolute zero has some **thermal energy**.

The **average thermal energy per particle** is (very roughly) kT.

k is Boltzmann's constant, $k = 1.38 \times 10^{-23}$ JK⁻¹. T is the temperature in kelvin.

2) This table gives you an idea of the magnitude of the thermal energy at various temperatures:

Temperature (K)	average thermal energy (approx.) — kT		
	J (per particle)	J mol⁻¹	eV (per particle)
1	1×10^{-23}	8	9×10^{-5}
300 (room temp)	4×10^{-21}	2000	0.03
6000 (Sun's surface)	8×10^{-20}	5×10^4	0.5

To convert kT to J mol⁻¹, multiply by Avogadro's constant (6.02×10^{23} particles per mole).

To convert kT to eV (electron-volts), divide by the charge on the electron (1.6×10^{-19} C).

3) Particles in matter are **held together** by **bonds**. The **energy** needed to break these bonds in a given substance is the **activation energy** ε (the Greek "epsilon").

4) The ratio ε/kT is really important. When kT is **big enough** compared with ε, the bonds are broken and the matter comes apart.

Lots of Processes have an **Activation Energy**

1) For a process like a change of state to happen, particles need to 'climb' an **energy barrier**.

2) The **activation energy**, ε, is the **energy needed** to climb that barrier (so, for a change of state, this activation energy is the latent heat — see p. 73).

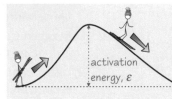

activation energy, ε

Before you can ski down a mountain, you need to climb to the top of it. So skiing is an activation process.

3) Lots of processes involving **particles** have activation energies — for example:

a) **A change of state**: the particles need enough energy to break the intermolecular forces.

b) **Thermionic emission**: if you heat up a conductor, electrons are released from the surface. These electrons need enough energy to escape from the attraction of the positive nuclei.

c) **Ionisation in a candle flame**: the molecules in the air need enough energy to split up into individual atoms, then ions. This is a similar process to thermionic emission.

d) **Conduction in a semiconductor**: semiconductors will only start to conduct once there are electrons in a high-energy state called the "conduction band", so electrons need enough energy to jump from the ground state to this higher-energy state.

4) In each of these examples, the **activation energy**, ε, comes from the **random thermal energy** of the particles. You might think, then, that these processes wouldn't happen unless $kT \geq \varepsilon$... but it's not that simple...

Getting **Extra Energy** is all about **Probabilities**

1) If the **ratio** between the activation energy and the average energy of the particles (ε/kT) is too high, nothing happens. But as ε/kT gets down to somewhere around **15–30**, the process starts to happen at a **fair rate**.

2) So some particles must have energies of **15–30 times** the **average energy**.

3) Every time particles **collide**, there's a **chance** that one of them will gain **extra energy** — above and beyond the average kT. If that happens to the **same particle** several times in a row (unlikely but possible), it can gain energies **much**, **much higher** than the average.

4) To end up with an energy of $15kT$ to $30kT$, a particle would have to get **very lucky**, so there will only be a tiny proportion of particles with this energy.

SECTION EIGHT — THERMAL PHYSICS & KINETIC THEORY

The Boltzmann Factor

The **Boltzmann Factor** tells you the **Ratio** of Particles in two Energy States

The **Boltzmann Factor** gives the **ratio** of the **numbers of particles** in energy states ε joules apart.

It's given by: $e^{-\frac{\varepsilon}{kT}}$

1) Processes start happening **quickly** when ε/kT is between 15 and 30, so try these values in the **Boltzmann factor**.

2) For $\varepsilon/kT = 15$, the Boltzmann factor is $\sim 10^{-7}$, and for $\varepsilon/kT = 30$ it's only $\sim 10^{-13}$.

3) That means that only **one in 10^{13}** to **one in 10^7** particles have **enough energy** to overcome the activation energy.

4) That might sound like a **tiny** proportion, but you have to remember how **fast** these particles are moving. Think about a reaction between two gases: gas particles collide about **10^9 times every second**. Every time there's a collision, there's an 'attempt' at the reaction, so even with **so few** particles having enough energy, the reaction can happen in a matter of **seconds**.

5) The **rate** of a reaction has quite a complicated relationship to **temperature**, but to a **reasonable approximation**:

The **rate** of a reaction with **activation energy** ε is proportional to the **Boltzmann factor**, $\exp(-\varepsilon/kT)$.

The **Boltzmann Factor** varies with **Temperature**

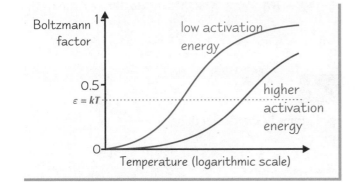

As long as the **thermal energy**, kT, is **lower** than the **activation energy**, ε, the Boltzmann factor **increases rapidly** with **temperature**.

So a **small increase** in **temperature** can make a **big difference** to the rate of a process or reaction.

Practice Questions

Q1 Give an expression for the approximate energy per particle at a given temperature.

Q2 The ratio ε/kT is important. What range of values must this ratio have, for processes to occur at a reasonable speed?

Q3 What is activation energy?

Q4 Explain how molecules can be ionised by a candle flame.

Exam Questions

Q1 A fish tank is in a room at a temperature of 300 K [$k = 1.38 \times 10^{-23}$ JK^{-1}].

  (a) Calculate the average energy of one of the water molecules in the tank. [1 mark]

  (b) Water molecules are joined together by two hydrogen bonds. The energy needed to break each bond is 3.2×10^{-20} J. Calculate the energy, ε, that a water molecule needs in order to evaporate. [1 mark]

  (c) Use your answers to parts (a) and (b) to find the ratio ε/kT. [1 mark]

  (d) Explain why the water in the tank must be topped up regularly. [3 marks]

The Boltzmann Factor — not as much fun as the Krypton Factor...

*You can think of the Boltzmann factor as the **probability** of a particle having a certain energy, or the fraction of particles that **do** have that energy. **Or** you could think of it as a big pair of grandma pants with pink polka dots — up to you...*

Probing to Determine Structure

This page is for AQA A, OCR A and OCR B only.

By firing radiation at different materials, you can take a sneaky beaky at what their internal structures look like...

Alpha Particle Scattering Lets Us See Inside the Atom...

1) If a beam of **positively charged alpha particles** is directed at a thin gold film, most **pass straight through**. However, if an alpha particle is travelling straight towards, or close by, a nucleus, its path will be **deflected**.

2) **Experimental evidence** shows that some of these alpha particles are deflected by **more than 90°** — in other words they '**bounce**' back.

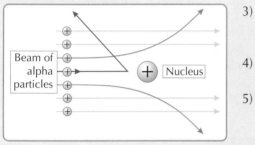

Rutherford Scattering

Beam of alpha particles

+ Nucleus

3) This evidence means that inside the atoms there must be **small positively charged nuclei** which **repel** the passing alpha particles.

4) The nucleus must be **small** since very few alpha particles are deflected by much.

5) It must be **positive** to repel the positively charged alpha particle.

You can estimate the **force** between an alpha particle and a nucleus of a gold atom using **Coulomb's law**. An alpha particle has a charge (Q_1) of +2e, and a gold nucleus (Q_2) of +79e. If the closest approach (r) between them is 1×10^{-13}m, then the force (F) is given by:

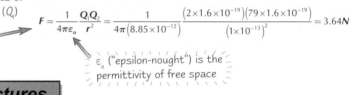

$$F = \frac{1}{4\pi\varepsilon_o}\frac{Q_1 Q_2}{r^2} = \frac{1}{4\pi(8.85\times10^{-12})}\frac{(2\times1.6\times10^{-19})(79\times1.6\times10^{-19})}{(1\times10^{-13})^2} = 3.64N$$

ε_o ("epsilon-nought") is the permittivity of free space

X-ray Diffraction Shows Crystal Structures

If a **beam of X-rays** is directed at a crystal, the **regularly arranged atoms** in the crystal diffract the beam in a particular way — creating a **diffraction pattern**.

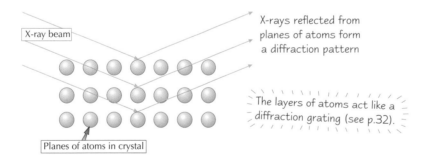

X-ray beam

X-rays reflected from planes of atoms form a diffraction pattern

The layers of atoms act like a diffraction grating (see p.32).

Planes of atoms in crystal

1) **X-ray diffraction** happens because the **wavelength** of X-rays is similar to the **atomic spacing** of the crystals.

2) At certain angles, the reflected X-rays **reinforce one another** to produce a **diffraction pattern**.

3) This diffraction pattern can be recorded using photographic paper and the **angles of diffraction** measured.

4) By **measuring these angles**, the **arrangement** and **spacing** of the atoms can be worked out.

Neutron Diffraction Gives a More Detailed Image of Crystal Structures

1) Neutrons are **particles**, but they also have **wave-like properties** (**wave-particle duality** — see p.64). Because of this, **neutrons** can be used to **produce diffraction patterns** for crystals, in a similar way to **X-rays**.

2) Neutrons also have a **magnetic moment**, which makes them **sensitive to magnetic forces** in materials.

3) These properties mean that beams of neutrons **interact more strongly** with nuclei than X-rays, and so give **more accurate information** about the crystal's structure.

Probing to Determine Structure

This page is for AQA A and OCR A only.

High-Energy Electrons *Give an* Accurate Picture *of the* Nucleus...

1) If you fire a beam of **high-energy electrons** at a material, some of them will be **scattered**, in much the same way as **Rutherford scattering**.

2) Electrons **are attracted** rather than **repelled** by the **electrostatic force** due to protons, and they're not **affected** by the **strong interaction** (see p. 82) — so electrons can get **a lot closer** to the nucleus than alpha particles.

3) This means that electron scattering produces a **more accurate picture** of nuclear structure than Rutherford scattering.

Electrons *can also be* Diffracted

1) Like other particles, electrons show **wave-particle duality** — so **electron beams** can be diffracted just like X-rays.

2) A beam of moving electrons has an associated **wavelength** that depends on the **voltage** the electron gun is operating at — the higher the voltage, the shorter the wavelength.

3) If a beam of **high-energy electrons** is directed onto a thin film of material, a **diffraction pattern** will be seen on the screen. Using measurements from this diffraction pattern, the **size and spacing** of the material's atomic **nuclei** can be worked out.

4) **Electrons** are a type of particle called a **lepton**, which **don't interact** with the **strong nuclear force** (whereas neutrons and alpha particles do). Because of this, electron diffraction is the **most accurate** method for getting a picture of a crystal's **atomic structure**.

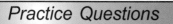

Practice Questions

Q1 Explain how alpha particle scattering shows that a nucleus is both small and positively charged.

Q2 Why are X-rays a suitable electromagnetic wave to investigate atomic sizes?

Q3 Neutrons have a magnetic moment. Why is this useful when investigating atomic structures?

Q4 How can the apparent wavelength of an electron beam be reduced?

Exam Questions

Q1 A beam of alpha particles is directed onto a very thin gold film.
(a) Explain why the majority of alpha particles are not scattered. [2 marks]
(b) Explain how alpha particles are scattered by atomic nuclei. [3 marks]

Q2 Various particles can be used to investigate the structure of matter.
(a) Why do particles such as neutrons or electrons produce diffraction patterns? [2 marks]
(b) Why is electron-beam diffraction the most accurate method for finding out about the atomic structure of a crystal? [1 mark]
(c) When electrons are directed at a larger nucleus, the beam suffers less diffraction. Why does this happen? [2 marks]

Alpha scattering — a bit like a game of miniature bowls...

Scattering and diffraction are the key ideas you need to understand for questions about atomic size and structure. Remember, particles like neutrons and electrons have wave-like properties, so if you fire them at crystal structures they make a diffraction pattern. This lets you work out size, spacing etc...

Nuclear Radius and Density

This page is for AQA A and OCR A (Core & Option 04) only.

The tiny nucleus, such a weird place, but one that you need to become ultra familiar with. Lucky you...

The **Nucleus** is a **Very Small Part** of a Whole **Atom**

1) By **probing atoms** using scattering and diffraction methods, we know that the **diameter of an atom** is about 0.1 nm (1×10^{-10} m) and the diameter of the smallest **nucleus** is about 2 fm (2×10^{-15} m — pronounced "femtometres").

2) So basically, **nuclei** are really, really **tiny** compared with the size of the **whole atom**.

3) To make this **easier to visualise**, try imagining a **large ferris wheel** (which is pretty darn big) as the size of **an atom**. If you then put a **grain of rice** (which is rather small) in the centre, this would be the size of the atom's **nucleus**.

4) **Molecules** are just a number of **atoms joined together**. As a rough guide, the size of a molecule equals the number of atoms in it multiplied by the size of one atom.

The **Nucleus** is **Made Up** of **Nucleons**

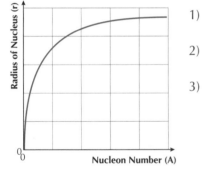

1) The **particles** that make up the nucleus (i.e. **protons** and **neutrons**) are called **nucleons**.

2) The **number of nucleons** in an atom is called the **mass (or nucleon) number, A**.

3) As **more nucleons** are added to the nucleus, it gets **bigger**.

See p. 86 for more on the mass number and how this is used to represent atomic structure in standard notation.

Nuclear Radius is **Proportional** to the **Cube Root** of the **Mass Number**

AQA A & OCR A Option 04

The **nuclear radius** increases roughly as the cube root of the mass (nucleon) number.

1) This **straight line graph** shows that the **nuclear radius** (r) is **directly proportional** to the cube root of the **nucleon number** (A).

2) This relationship can be written as: $r \propto A^{1/3}$.

3) We can make this into an equation by introducing a constant, r_o, giving:

$$r = r_o A^{1/3}$$

Where r_o is the value of r when $A = 1$ i.e. for a proton (hydrogen nucleus). The value of r_o is about 1.4 fm.

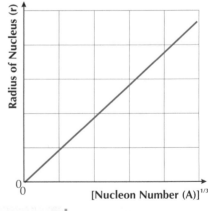

Example   Calculate the radius of an oxygen nucleus which has 16 nucleons.

$$r = r_o A^{1/3} = 1.4 \times 10^{-15} \times (16)^{1/3}$$
$$= 3.5 \times 10^{-15} \text{ m (or 3.5 fm)}$$

Nuclear Radius and Density

This page is for OCR A Option 04 *only.*

The **Density** of **Nuclear** Matter is **Enormous**

1) The **volume** that each nucleon (i.e. a **proton** or a **neutron**) takes up in a nucleus is about the **same**.

2) Because protons and neutrons have nearly the **same mass**, it means that all nuclei have a **similar density** (ρ).

3) But nuclear matter is **no ordinary** stuff. Its density is **enormous**. A **teaspoon** of pure nuclear matter would have a mass of **five hundred million tonnes**. (Just to make you gasp in awe and wonder, out in space nuclear matter makes up neutron stars which are several kilometres in diameter.)

The following examples show how **nuclear density** is pretty much the **same**, **regardless of the element**.

Example 1
Work out the density of a carbon nucleus given that its mass is 2.00×10^{-27} kg.

1) The **radius** (r) of a carbon nucleus $\approx 3.2 \times 10^{-15}$ m

2) So, the **volume** (v) of the nucleus $= \frac{4}{3}\pi r^3$
$$= 1.37 \times 10^{-43} \text{ m}^3$$

3) This gives the **density** (ρ) of a carbon nucleus as:
$$\rho = \frac{m}{v} = \frac{2.00 \times 10^{-26}}{1.37 \times 10^{-43}} = 1.46 \times 10^{17} \text{ kg m}^{-3}$$

Example 2
Work out the density of a gold nucleus given that its mass is 3.27×10^{-25} kg.

1) The **radius** (r) of a gold nucleus $\approx 8.1 \times 10^{-15}$ m

2) So, the **volume** (v) of the nucleus $= \frac{4}{3}\pi r^3$
$$= 2.23 \times 10^{-42} \text{ m}^3$$

3) This gives the **density** (ρ) of a gold nucleus as:
$$\rho = \frac{m}{v} = \frac{3.27 \times 10^{-25}}{2.22 \times 10^{-42}} = 1.47 \times 10^{17} \text{ kg m}^{-3}$$

Nuclear density is significantly larger than atomic density — this suggests three important facts about the structure of an atom:
a) Most of an atom's mass is in its nucleus.
b) The nucleus is small compared to the atom.
c) An atom must contain a lot of empty space.

Practice Questions

Q1 What is the approximate size of an atom?

Q2 What are nucleons?

Q3 What is the relationship between the nuclear radius and mass number?

Q4 In the formula $r = r_o A^{1/3}$, what does r_o represent?

Q5 Explain why the density of ordinary matter is much less than that of nuclear matter.

Exam Questions

Q1 The radius (r) of a nucleus with A nucleons can be calculated using the equation $r = r_o A^{1/3}$.
 (a) If a carbon nucleus containing 12 nucleons has a radius of 3.2×10^{-15} m, show that $r_o = 1.4 \times 10^{-15}$ m. [2 marks]
 (b) Calculate the radius of a radium nucleus containing 226 nucleons. [1 mark]
 (c) Calculate the density of the radium nucleus if its mass is 3.75×10^{-25} kg. [3 marks]

Q2 A sample of pure gold has a density of 19300 kgm⁻³. If the density of a gold nucleus is 1.47×10^{17} kgm⁻³, discuss what this implies about the structure of a gold atom. [4 marks]

Nuclear and particle physics — heavy stuff...

So basically the nucleus is a tiny part of the atom, but it's incredibly dense. The density doesn't change from element to element, and the radius depends on the mass number. Learn the theory like your own back yard, but don't worry too much about equations and values — those friendly examiners have popped them in the exam paper for you. How nice...

The Strong Interaction

*These pages are for **AQA B** and **OCR A Option 04** only.*

Keeping the nucleus together requires a lot of effort — a bit like A2 Physics then...

There are **Forces** at Work **Inside** the **Nucleus**...

There are several different **forces** acting on the nucleons in a nucleus. To understand these forces you need to take a look at **electrostatic** forces due to the protons' electric charges, and also the **gravitational** forces.

1) The **electrostatic force**

All protons have an equal, **positive electric charge**. So, packed close together inside a nucleus, these protons will **repel** each other. We can work out the size of this **force of repulsion** using **Coulomb's Law**. Assuming that two protons are 1×10^{-14} m apart, the force of repulsion F_R can be given as:

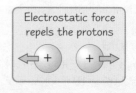

Electrostatic force repels the protons

$$F_R = \frac{1}{4\pi\varepsilon_o} \frac{Q_1 Q_2}{r^2} = \frac{1}{4\pi(8.85\times10^{-12})} \frac{(1.6\times10^{-19})(1.6\times10^{-19})}{(1\times10^{-14})^2} = 2.3 \text{ N}$$

In this example: Q_1 and Q_2 are the electric charges on two protons, ε_o is the permittivity of free space, and r is the distance separating the two protons.

2) The **gravitational force**

Newton's law of gravitation says that two **massive** objects will **attract** each other. So, for two protons in a nucleus, this **attractive force**, F_A, can be given as:

Gravitational force attracts the protons

$$F_A = -G \frac{m_1 m_2}{r^2} = -(6.67\times10^{-11}) \cdot \frac{(1.67\times10^{-27})(1.67\times10^{-27})}{(1\times10^{-14})^2} = -1.86 \times 10^{-36} \text{ N}$$

In this example: G is the gravitational constant, m_1 and m_2 are the masses of two protons, and r is the distance separating the two protons.

The **electrostatic force** of repulsion is far **bigger** than the **gravitational** attractive force. If these were the only forces acting in the nucleus, the nucleons would **fly apart**. So there must be **another attractive force** that **holds the nucleus together** — called the **strong interaction** or **strong nuclear force**. (The gravitational force is so small, it's usually ignored.)

AQA B: the strong nuclear force acts between all **quarks** and **hadrons**, not just nucleons. (You covered quarks and hadrons at AS, but have a look at pages 96-101 if you need a quick reminder.)

The **Strong Interaction** Binds Nucleons Together

Now pay attention please. This bit's rather strange because the **strong interaction** is quite **complicated**, but here are the **main points**:

1) To **hold the nucleus together**, the strong interaction must be an **attractive force** that is **larger** than the electrostatic force.

2) Experiments have shown that the strong interaction has a **short range**. It can only hold nucleons together when they are separated by up to **10 fm** — which is the maximum size of a nucleus.

3) The **strength** of the strong interaction **quickly falls** beyond this distance (see the graph on the next page).

4) Experiments also show that the strong interaction **works equally between all nucleons**. This means that the size of the force is the same whether proton-proton, neutron-neutron or proton-neutron.

5) At **very small separations**, the strong interaction must be **repulsive** — otherwise there would be nothing to stop it **crushing** the nucleus to a **point**.

lime green, orange and day-glow pink — repulsive at small separations

The Strong Interaction

The Size of the Strong Interaction Varies with Nucleon Separation

The **strong interaction force** can be plotted on a **graph** to show how it changes with the **distance of separation** between **nucleons**. If the **electrostatic force** is also plotted the **relationship** between these **two forces** can be seen.

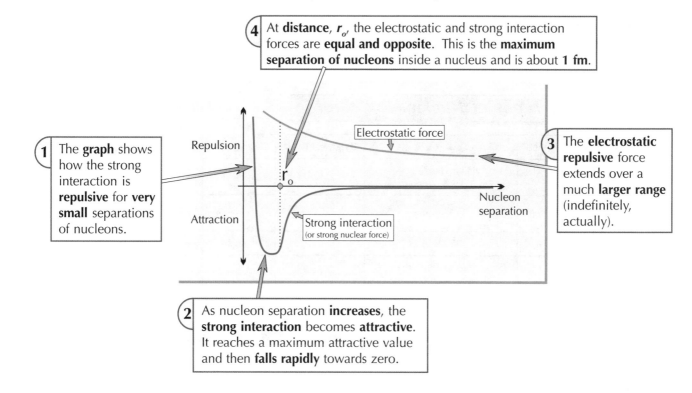

4 At **distance**, r_o, the electrostatic and strong interaction forces are **equal and opposite**. This is the **maximum separation of nucleons** inside a nucleus and is about **1 fm**.

1 The **graph** shows how the strong interaction is **repulsive** for **very small** separations of nucleons.

3 The **electrostatic repulsive** force extends over a much **larger range** (indefinitely, actually).

2 As nucleon separation **increases**, the **strong interaction** becomes **attractive**. It reaches a maximum attractive value and then **falls rapidly** towards zero.

Practice Questions

Q1   What causes an electrostatic force inside the nucleus?

Q2   Explain why the gravitational forces between nucleons are usually ignored.

Q3   What evidence suggests the existence of a strong nuclear force?

Q4   Is the strong interaction attractive or repulsive at a nucleon separation of 10 fm?

Exam Questions

Q1   Coulomb's law can be used to find the electrostatic force of repulsion between two protons in a nucleus.

$$F = \frac{1}{4\pi\varepsilon_o} \frac{Q_1 Q_2}{r^2}$$

The charge, Q, on a proton is $+1.6 \times 10^{-19}$ C and the permittivity of free space, ε_0, is 8.85×10^{-12} Fm^{-1}.

(a) If two protons are separated by a distance, r, of 9×10^{-15} m, calculate the electrostatic force between them.   [2 marks]

(b) If the protons move closer together, what effect will this have on the repulsive force?   [1 mark]

(c) What is the electrostatic force between a proton and a neutron?   [2 marks]

Q2   The strong nuclear force binds the nucleus together.

(a) Explain why the force must be repulsive at very short distances.   [1 mark]

(b) How does the strong interaction limit the size of a stable nucleus?   [2 marks]

The strong interaction's like nuclear glue...

Right then, lots of scary looking stuff on these pages, but DON'T PANIC... the important bits can be condensed into a few points: a) the electrostatic force pushes protons in the nucleus apart, b) the strong interaction pulls all nucleons together, c) there's a point where these forces are balanced — this is the maximum nucleon separation in a nucleus. Easy eh?...

Radioactive Emissions

These pages are for AQA A, OCR A (you only need to know about β⁺ if you're doing Option 04) and OCR B only.

Unstable Atoms are Radioactive

1) If an atom is **unstable**, it will **break down** to **become** more stable. This **instability** could be caused by having **too many neutrons**, **not enough neutrons**, or just **too much energy** in the nucleus.

2) The atom **decays** by **releasing energy** and/or **particles**, until it reaches a **stable form** — this is called **radioactive decay**.

3) Radioactivity is **random** — it can't be predicted (see p. 88).

There are Four Types of Nuclear Radiation

Learn this table.

a.m.u. stands for atomic mass unit.

See p. 86 for more on positrons.

Radiation	Symbol	Constituent	Relative Charge	Mass (a.m.u.)
Alpha	α	A helium nucleus — 2 protons & 2 neutrons	+2	4
Beta-minus (Beta)	β or β⁻	Electron	-1	(negligible)
Beta-plus	β⁺	Positron	+1	(negligible)
Gamma	γ	Short-wave, high-frequency electromagnetic wave.	0	0

The Different Types of Radiation have Different Penetrations

You've got to learn this table as well.

When a radioactive particle **hits** an **atom** it can **knock off** the **outer electrons**, creating an **ion** — so, **radioactive emissions** are also known as **ionising radiation**. **Alpha**, **beta** and **gamma** radiation can be **fired** at a **variety of objects** with **detectors** placed the **other side** to see whether they **penetrate** the object.

Skin or paper stops ALPHA

Many cm lead stops GAMMA

Thin mica

Few mm aluminium stops BETA

Radiation	Symbol	Ionising	Speed	Penetrating power	Affected by magnetic field
Alpha	α	Strongly	Slow	Absorbed by paper or a few cm of air	Yes
Beta-minus (Beta)	β or β⁻	Weakly	Fast	Absorbed by ~3 mm of aluminium	Yes
Beta-plus	β⁺	Annihilated by electron — so virtually zero range			
Gamma	γ	Very weakly	Speed of light	Absorbed by many cm of lead, or several m of concrete.	No

Radiation can Cause a lot of Harm to Body Tissues   *OCR B only.*

1) The amount of **energy absorbed per kilogram** of tissue is called the **absorbed dose**, and is measured in **grays (Gy)**.

2) But the amount of **tissue damage** isn't just due to the amount of energy absorbed — it also depends on the **type of ionising radiation** and the **type of body tissue**.

3) The **dose equivalent** is a measure that lets us **compare the amount of damage** to body tissues that have been **exposed** to different types of radiation:

See p.162 for more on the sievert and biological effects of radiation.

Dose equivalent = absorbed dose × radiation quality factor

4) The **unit** of dose equivalence is the **sievert (Sv)**.

5) **For example**, if you exposed a sample of **body tissue** to **1 Gy** of **alpha** radiation, it would do the **same damage** as an exposure of 20 Gy of **gamma** radiation on the same type of body tissue.

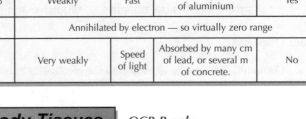

Radiation	Radiation quality factor	Dose equivalent of 1 Gy
alpha	20	20 Sv
beta	1	1 Sv
gamma	1	1 Sv

Radioactive Emissions

Alpha and *Beta* Particles have *Different Ionising Properties*

The **different dose equivalent** for each type of radiation can be **explained** by their **ionising properties**.

1) **Alpha** particles are **strongly positive** — so they can **easily pull electrons** off atoms.

2) Ionising an atom **transfers** some of the **energy** from the **alpha particle** to the **atom**. The alpha particle **quickly ionises** many atoms (about 10 000 ionisations per alpha particle) and **loses** all its **energy** — that's why it causes so much **damage** to body tissue and is given a **high dose equivalent** value.

3) The **beta**-minus particle has **lower charge** than the alpha particle, but a **higher speed**. This means it can still **knock electrons** off atoms. Each **beta** particle will ionise about 100 atoms, **losing energy** at each interaction.

4) This **lower interaction** rate means that beta radiation causes much **less damage** to body tissue than alpha radiation — explaining the **lower dose equivalent** value.

The *Intensity* of *Gamma Radiation* Obeys the *Inverse Square Law*

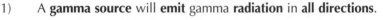

1) A **gamma source** will **emit** gamma **radiation** in **all directions**.

2) This radiation **spreads out** as you get **further away** from the source.

3) However, the amount of **radiation per unit area** (the **intensity**) will **decrease** the further you get from the source.

4) If you took a reading of **intensity**, *I*, at a **distance**, *x*, from the source you would find that it **decreases** by the **square of the distance** from the source.

5) This can be written as the equation:

$$I = \frac{kI_o}{x^2}$$ where k is a constant, and I_o is the intensity at the source.

6) This **relationship** can be **proved** by taking **measurements of intensity** at different distances from a gamma source, using a **Geiger-Müller tube** and **counter**.

7) If the **distance** from the source is **doubled** the **intensity** is found to **fall to a quarter** — which **verifies** the inverse square law.

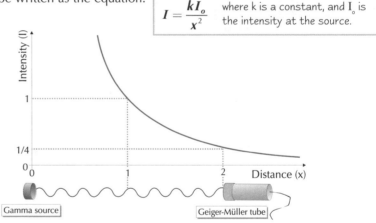

Practice Questions

Q1 What makes an atom radioactive?

Q2 Name three types of nuclear radiation and give three properties of each.

Q3 What does the term 'dose equivalent' mean?

Exam Questions

Q1 Draw a diagram showing the relative penetration of alpha, beta and gamma radiation. [4 marks]

Q2 Which of the following doses would cause the greatest damage to a sample of body tissue:
0.6 Gy of alpha radiation or 9 Gy of beta radiation? [3 marks]

Q3 The count rate detected by a GM-tube, 10 cm from a gamma source, is 240 counts per second.
What would you expect the count rate to be at 40 cm from the source? [3 marks]

Radioactive emissions — as easy as α, β, γ...

You need to learn the different types of radiation and their properties. Remember that alpha particles are by far the most ionising and so cause more damage to body tissue than the same dose of any other radiation — which is one reason we don't use alpha sources as medical tracers. Learn this all really well, then go and have a brew and a bickie...

Nuclear Decay

This page is for AQA A, OCR A and OCR B only.

The stuff on these pages covers the most important facts about nuclear decay that you're just going to have to make sure you know inside out. I'd be very surprised if you didn't get a question about it in your exam...

Atomic Structure can be Represented Using Standard Notation

STANDARD NOTATION:

The nucleon number or mass number (A) — there are a total of 12 protons and neutrons in a carbon-12 atom.

The proton number or atomic number (Z) — there are six protons in a carbon atom.

$$^{12}_{6}\text{C}$$

The symbol for the element carbon.

Atoms with the **same number of protons** but **different numbers of neutrons** are called **isotopes**. The following examples are all isotopes of carbon: $^{12}_{6}\text{C}$, $^{13}_{6}\text{C}$, $^{14}_{6}\text{C}$

Some Nuclei are More Stable than Others

The nucleus is under the **influence** of the **strong nuclear force holding** it **together** and the **electromagnetic force pushing** the **protons apart**. It's a very **delicate balance**, and it's easy for a nucleus to become **unstable**. You can get a stability graph by plotting **Z** (atomic number) against **N** (number of neutrons).

A nucleus will be **unstable** if it has:

1) **too many neutrons**

2) **too few neutrons**

3) **too many nucleons** altogether, i.e. it's **too heavy**

4) **too much energy**

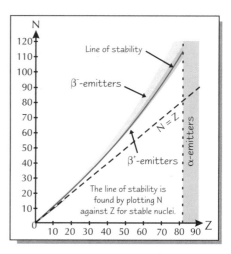

α Emission Happens in Heavy Nuclei

When an alpha particle is **emitted**:

The **proton number decreases** by **two**, and the **nucleon number decreases** by **four**.

nucleon number decreases by 4

$$^{238}_{92}\text{U} \longrightarrow {}^{234}_{90}\text{Th} + {}^{4}_{2}\alpha$$

proton number decreases by 2

1) **Alpha emission** only happens in **very heavy** atoms (with more than 82 protons), like **uranium** and **radium**.

2) The **nuclei** of these atoms are **too massive** to be stable.

β⁻ Emission Happens in Neutron Rich Nuclei

1) **Beta-minus** (usually just called beta) decay is the emission of an **electron** from the **nucleus** along with an **antineutrino**.

2) Beta decay happens in isotopes that are **"neutron rich"** (i.e. have many more **neutrons** than **protons** in their nucleus).

3) When a nucleus ejects a beta particle, one of the **neutrons** in the nucleus is **changed** into a **proton**.

4) In **beta-plus emission**, a **proton** gets **changed** into a **neutron**. The **proton number decreases** by **one**, and the **nucleon number stays the same**.

When a beta particle is **emitted**:

The **proton number increases** by **one**, and the **nucleon number stays the same**.

nucleon number stays the same

$$^{187}_{75}\text{Re} \longrightarrow {}^{187}_{76}\text{Os} + {}^{0}_{-1}\beta$$

proton number increases by 1

Nuclear Decay

This page is for AQA A, OCR A, OCR B and AQA B (just the conservation rules) only.

γ **Radiation** is Emitted from **Nuclei** with **Too Much Energy**

1) **After alpha** or **beta** decay, the **nucleus** often has **excess energy** — this energy is **lost** by emitting a **gamma ray**.

2) **Another way** that gamma radiation is produced is when a nucleus **captures** one of its own orbiting **electrons**.

3) **Electron capture** causes a **proton** to **change** into a **neutron**. This makes the **nucleus unstable** and it **emits** gamma radiation.

4) **Gamma rays** can also be emitted from stable, yet **excited nuclei** — as they fall to a **lower energy** level they **emit** a gamma ray.

> During gamma emission, there is **no change** to the nuclear **constituents** — the nucleus just **loses excess energy**.

> The artificial isotope technetium-99^m is formed in an excited state from the decay of another element. It is used as a tracer in medical imaging.

There are **Conservation Rules** in **Nuclear Reactions**

In every nuclear reaction **energy**, **momentum**, **proton number** and **nucleon number** must be conserved.

234 + 4 = 238 — nucleon numbers balance

$$^{238}_{92}U \longrightarrow \, ^{234}_{90}Th + \, ^{4}_{2}\alpha$$

90 + 2 = 92 — proton numbers balance

Mass is **Not Conserved**

1) The **mass** of the **alpha particle** is less than the **individual masses** of **two protons** and **two neutrons**. The difference is called the **mass defect**.

2) Mass **doesn't** have to be **conserved** because of **Einstein's equation**:

$$E = mc^2$$

3) This says that **mass and energy** are **equivalent**. The **energy released** when the nucleons **bonded together** accounts for the missing mass — so the **energy released** is the same as the **mass defect × c^2**.

Practice Questions

Q1 What makes a nucleus unstable?

Q2 Describe the changes that happen in the nucleus during alpha, beta and gamma decay.

Q3 Explain the circumstances in which gamma radiation may be emitted.

Q4 Define the mass defect.

Exam Questions

Q1 (a) Radium-226 undergoes alpha decay to radon. Complete the balanced nuclear equation for this reaction. [3 marks]

$$^{226}_{88}Ra \rightarrow \, Rn +$$

(b) Potassium-40 ($Z = 19$, $A = 40$) undergoes beta decay to calcium.
Write a balanced nuclear equation for this reaction. [3 marks]

Q2 Calculate the energy released during the formation of an alpha particle, given that the total mass of two protons and two neutrons is 6.693×10^{-27} kg, the mass of an alpha particle is 6.425×10^{-27} kg and the speed of light, c, is 3.00×10^8 ms^{-2}. [3 marks]

Einstein's equation explains where the missing mass goes...

$E = mc^2$ is an important equation that says mass and energy are equivalent. Remember it well, 'cos you're going to come across it a lot in questions about mass defect and the energy released in nuclear reactions over the next few pages...

Exponential Law of Decay

These pages are for AQA A, AQA B, OCR A and OCR B only.

Oooh look — some maths. Good.

Every Isotope *Decays* at a *Different Rate*

1)   **Radioactive decay** is completely **random**. You **can't predict which** atom will decay **when**.

2)   Although you can't predict the decay of an **individual atom**, if you take a **very large number of atoms**, their **overall behaviour** shows a **pattern**.

3)   Any sample of a particular **isotope** has the **same rate of decay**, i.e. the same **proportion** of atoms will **decay** in a **given time**.

It could be you.

The *Rate of Decay* is Measured by the *Decay Constant*

The **activity** of a sample — the **number** of atoms that **decay each second** — is **proportional** to the **size of the sample**. For a **given isotope**, a sample **twice** as big would give **twice** the **number of decays** per second.

The **decay constant** (λ) measures how **quickly** something will **decay** — the **bigger** the value of λ, the faster the rate of decay. Its unit is s^{-1}.

> activity = decay constant × number of atoms

Or in symbols: $A = \lambda N$ ← Don't get λ confused with wavelength.

Activity is measured in **becquerels** (Bq):

> 1 Bq = 1 decay per second (s^{-1})

You Need to *Learn* the *Definition of Half-Life*

> The **half-life** ($T_{1/2}$) of an **element** is the **average time** it takes for the **number of undecayed atoms** to **halve**.

Measuring the **number of undecayed atoms** isn't the easiest job in the world. **In practice**, half-life isn't measured by counting atoms, but by measuring the **time it takes** the **activity** to **halve**.

The **longer** the **half-life** of an isotope, the **longer** it stays **radioactive**.

The *Number* of *Undecayed* Particles *Decreases Exponentially*

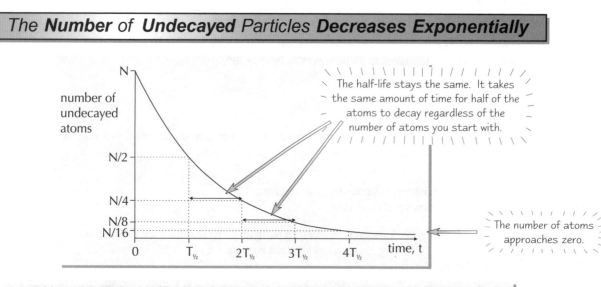

The half-life stays the same. It takes the same amount of time for half of the atoms to decay regardless of the number of atoms you start with.

The number of atoms approaches zero.

You'd be **more likely** to actually meet a **count rate-time graph** or an **activity-time graph**. They're both **exactly the same shape** as the graph above, but with different **y-axes**.

Exponential Law of Decay

You can *Find* the *Half-Life* of an *Isotope* from the *Graph*

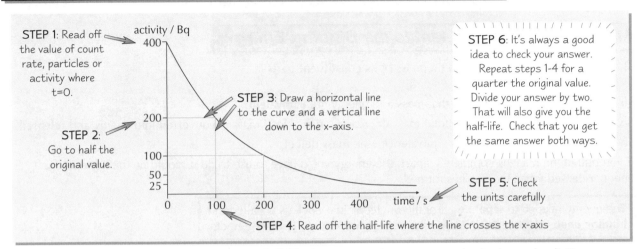

STEP 1: Read off the value of count rate, particles or activity where t=0.

STEP 2: Go to half the original value.

STEP 3: Draw a horizontal line to the curve and a vertical line down to the x-axis.

STEP 6: It's always a good idea to check your answer. Repeat steps 1-4 for a quarter the original value. Divide your answer by two. That will also give you the half-life. Check that you get the same answer both ways.

STEP 5: Check the units carefully

STEP 4: Read off the half-life where the line crosses the x-axis

When you're **measuring** the **activity** and **half-life** of a **source**, you've got to **remember background radiation**. The **background radiation** needs to be **subtracted** from the **activity readings** to give the **source activity**.

You Need to Know the *Equations* for *Half-Life* and *Decay*...

1) The **half-life** can be **calculated** using the equation:
(where ln is the natural log)

$$T_{\frac{1}{2}} = \frac{\ln 2}{\lambda}$$

2) The **number of radioactive atoms** remaining, *N*, depends on the **number originally** present, N_o. The **radioactive decay** can be calculated using the equation:

$$N = N_0 e^{-\lambda t}$$

Here *t* = time, measured in seconds.

Example:
A sample of the radioactive isotope ^{13}N contains 5×10^6 atoms. The decay constant for this isotope is 1.16×10^{-3} s^{-1}.

a) What is the half-life for this isotope?

$$T_{\frac{1}{2}} = \frac{\ln 2}{1.16 \times 10^{-3}} = 598 \text{ s}$$

b) How many atoms of ^{13}N will remain after 800 seconds?

$$N = N_0 e^{-\lambda t} = 5 \times 10^6 e^{-(1.16 \times 10^{-3})(800)} = 1.98 \times 10^6 \text{ atoms}$$

Practice Questions

Q1 Define radioactive activity. What units is it measured in?

Q2 Sketch a general radioactive decay graph showing the number of particles against time.

Q3 What is meant by the term 'half-life'? Give two methods for working out the half-life of a radioactive sample.

Exam Questions

Q1 You take a reading of 750 Bq from a pure radioactive source. The radioactive source initially contains 50 000 atoms, and background activity in your lab is measured as 50 Bq.
(a) Calculate the decay constant for your sample. [3 marks]
(b) What is the half-life of this sample? [2 marks]
(c) Approximately how many atoms of the radioactive source will there be after 300 seconds? [2 marks]

Radioactivity is a random process — just like revision shouldn't be...

Remember the shape of that graph — whether it's count rate, activity or number of atoms plotted against time, the shape's always the same. This is all pretty straightforward mathsy-type stuff: plugging values in equations, reading off graphs, etc. Not very interesting, though. Ah well, once you get onto relativity you'll be longing for a bit of boredom.

Binding Energy

This page is for AQA A, AQA B, OCR A Option 04 and OCR B only.

Turn off the radio and close the door, 'cos you're going to need to concentrate hard on this stuff about binding energy...

The **Mass Defect** is **Equivalent** to the **Binding Energy**

1) The **mass** of a **nucleus** is **less than** the mass of its **constituent parts**
    — this is called the **mass defect** (see p. 87).

2) Einstein's equation, $E = mc^2$, says that mass and energy are **equivalent**.

3) So, as nucleons join together, the total mass **decreases** — this 'lost' mass is **converted** into energy and **released**.

4) The amount of **energy released** is **equivalent** to the **mass defect**.

5) If you **pulled** the nucleus completely **apart**, the **energy** you'd have to use to do it would be the **same** as the energy **released** when the nucleus formed.

> The energy needed to **separate** all of the nucleons in a nucleus is called the
> **binding energy** (measured in **MeV**), and it is **equivalent** to the **mass defect**.

Atomic mass is usually given in atomic mass units (u), where $1\,u = 1.66 \times 10^{-27}$ kg.

Example Calculate the binding energy of the nucleus of a lithium atom, $^{6}_{3}\text{Li}$, given that its mass defect is 0.034346 u.

  1) Convert the mass defect into kg.
      Mass defect = $0.034346 \times 1.66 \times 10^{-27} = 5.70 \times 10^{-29}$ kg

  2) Using $E = mc^2$ to calculate the binding energy.
      $E = 5.70 \times 10^{-29} \times (3 \times 10^8)^2 = 5.13 \times 10^{-12}$ J = 32 MeV

$1\,\text{MeV} = 1.6 \times 10^{-13}$ J

6) The **binding energy per unit of mass defect** can be calculated (using the example above):

$$\frac{\text{Binding energy}}{\text{mass defect}} = \frac{32\ \text{MeV}}{0.034346\ \text{u}} \approx 931\ \text{MeV}\,\text{u}^{-1}$$

7) This means that a mass defect of **1 u ≈ 931 MeV** of binding energy.

The **Binding Energy Per Nucleon** is at a **Maximum** around **N = 50**

A useful way of **comparing** the binding energies of different nuclei is to look at the **binding energy per nucleon**.

> Binding energy per nucleon (in MeV) = $\dfrac{\text{Binding energy (B)}}{\text{Nucleon number (A)}}$

So, the binding energy per nucleon for $^{6}_{3}\text{Li}$
(in the example above) is 32 ÷ 6 = 5.3 MeV.

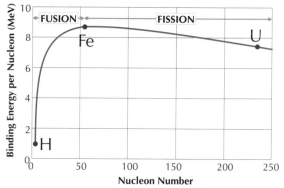

1) A **graph** of **binding energy per nucleon** against **nucleon number**, for all elements, shows a **curve**.

2) **High** binding energy per nucleon means that **more energy** is needed to **remove** nucleons from the nucleus.

3) In other words the **most stable** nuclei occur at the **maximum point** on the graph
    — which is at about **nucleon number 56** (i.e. **Iron**, Fe).

4) **Combining small nuclei** is called nuclear **fusion** (see p. 94) — this **increases** the **binding energy per nucleon** dramatically, which means a lot of **energy is released** during nuclear fusion.

5) **Fission** is when **large nuclei** are **split in two** (see p. 92) — the **nucleon numbers** of the two **new nuclei** are **smaller** than the original nucleus, which means there is an **increase** in the binding energy per nucleon. So, energy is also **released** during nuclear fission (but not as much energy per nucleon as in nuclear fusion).

Binding Energy

This page is for OCR A Option 04 and OCR B only.

The Change in Binding Energy Gives the Energy Released...

The **binding energy per nucleon graph** can be used to **estimate** the **energy released** from nuclear reactions.

Energy released in nuclear fusion

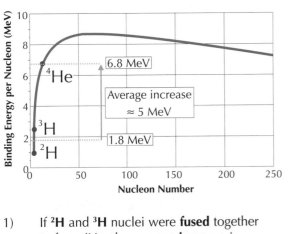

1) If **²H** and **³H** nuclei were **fused** together to form ⁴He, the **average increase** in binding energy per ⁴He nucleon would be about **5 MeV**.

2) There are **4 nucleons** in ⁴He, so we can **estimate** the **energy released** as 4 × 5 = **20 MeV**.

Energy released in nuclear fission

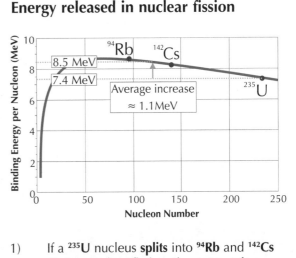

1) If a **²³⁵U** nucleus **splits** into **⁹⁴Rb** and **¹⁴²Cs** during nuclear **fission**, the **average increase** in **binding energy per nucleon** would be about 1.1 MeV.

2) There are **235 nucleons** in ²³⁵U to begin with, so we can **estimate** the energy **released** as 235 × 1.1 ≈ **260 MeV**.

Practice Questions

Q1 What is the binding energy of a nucleus?

Q2 How can we calculate the binding energy for a particular nucleus?

Q3 What is the binding energy per nucleon?

Q4 Which element has the highest value of binding energy per nucleon?

Q5 Do nuclear fusion or fission reactions release the most energy per nucleon?

Exam Questions

Q1 The mass of a $^{14}_{6}C$ nucleus is 13.999948 u. The mass of a proton is 1.007276 u, and a neutron is 1.008665 u.

(a) What is the mass defect of a $^{14}_{6}C$ nucleus (given that 1 u = 1.66×10^{-27} kg)? [3 marks]

(b) Use $E = mc^2$ to calculate the binding energy of the nucleus in MeV
(given that c = 3×10^8 ms⁻² and 1 MeV = 1.6×10^{-13} J). [2 marks]

Q2 The following equation represents a nuclear reaction that takes place in the Sun:

$$^1_1p + {}^1_1p \rightarrow {}^2_1H + {}^0_{+1}\beta + \text{energy released}$$ where p is a proton and β is a positron (opposite of an electron)

(a) What type of nuclear reaction is this? [1 mark]

(b) Given that the binding energy per nucleon for a proton is 0 MeV and for a ²H nucleus it is approximately 0.86 MeV, estimate the energy released by this reaction. [2 marks]

A mass defect of 1 u is equivalent to a binding energy of 931 MeV...

Remember this useful little fact, and it'll save loads of time in the exam — because you don't have to fiddle around with converting atomic mass from u → kg and binding energy from J → MeV. What more could you possibly want...

Nuclear Fission

These pages are for AQA A, AQA B, OCR A and OCR B only.

Fission Means Splitting Up Into Smaller Parts

1) **Large nuclei**, with more than 92 protons (e.g. uranium), are **unstable** and some can randomly **split** into two **smaller** nuclei — this is called **nuclear fission**.

2) This process is called **spontaneous** if it just happens **by itself** or **induced** if we **encourage** it to happen.

3) It might help to imagine **large** unstable **nuclei** as drops of liquid — this is called the **liquid-drop model**.

LIQUID-DROP MODEL

1) If the drop is too big, there won't be enough force to hold the drop together, so it splits into two drops all by itself (spontaneous fission).
2) If you fire something at the drop, it'll make it oscillate. If the oscillations are big enough, the drop splits into two separate droplets (induced fission).

Example

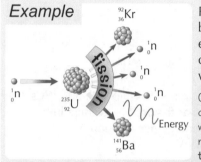

$^{92}_{36}Kr$

$^{1}_{0}n$

$^{1}_{0}n$

$^{1}_{0}n$

$^{1}_{0}n$

$^{235}_{92}U$

fission

Energy

$^{141}_{56}Ba$

Fission can be induced by making a neutron enter a ^{235}U nucleus, causing it to become very unstable.

Only low energy neutrons can be captured in this way. A low energy neutron is called a thermal neutron.

4) **Energy is released** during nuclear fission because the new, smaller nuclei have a **higher binding energy per nucleon** (see p. 78).

5) The **larger** the nucleus, the more **unstable** it will be — so large nuclei are **more likely** to **spontaneously fission**.

6) This means that spontaneous fission **limits** the **number of nucleons** that a nucleus can contain — or in other words, it **limits** the number of **possible elements**.

There are Different Possible Fission Products

1) The example above shows the **fission reaction**:

$$^{235}_{92}U + ^{1}_{0}n \rightarrow ^{92}_{36}Kr + ^{141}_{56}Ba + 3^{1}_{0}n + \text{energy released}$$

2) But this is just **one way** that the uranium nucleus could have split — there are many other **different fission products** that could have been made.

fish'n' products... ho ho ho ho

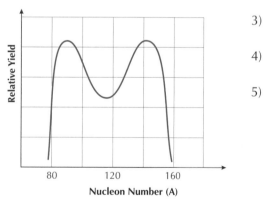

Relative Yield

80   120   160

Nucleon Number (A)

3) This **graph** shows the **spread** of possible fission products produced by splitting a ^{235}U nucleus.

4) The most common fragments can be seen at the graph's peaks — i.e. with **atomic masses around 90 and 140**.

5) In any fission, the products must '**add up**' to give the **same number of protons and neutrons** as at the start.

For example, in the fission reaction given above, the total number of protons and neutrons before the reaction (235 + 1 = 236) must equal the total after (92 + 141 + 3 = 236). Similarly, the total number of protons before the reaction (92 + 0 = 92) must equal the total after (36 + 56 = 92).

Fission Products can be Highly Radioactive

1) The **fission products** usually have a **larger proportion of neutrons** than nuclei of a similar atomic number — this makes them **unstable** and **radioactive**.

2) The products can be used for **practical applications** such as **tracers** in medical diagnosis.

3) However the products may be **highly radioactive** and so their **handling** and **disposal** needs **great care**.

Nuclear Fission

Controlled **Nuclear Reactors** Produce Useful **Power**

We can **harness** the **energy** released during nuclear **fission reactions** in a **nuclear reactor**, but it's important that these reactions are very **carefully controlled**.

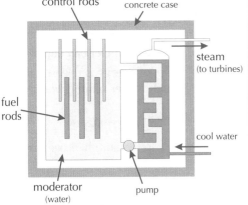

1) Nuclear reactors use **rods of uranium** that are rich in ^{235}U as 'fuel' for fission reactions. (The rods also contain a lot of ^{238}U, but that doesn't undergo fission — see the "plutonium" bit, below.)

2) These **fission** reactions produce more **neutrons** which then **induce** other nuclei to fission — this is called a **chain reaction**.

3) The **neutrons** will only cause a chain reaction if they are **slowed down**, which allows them to be **captured** by the uranium nuclei — these slowed down neutrons are called **thermal neutrons**.

4) To do this the ^{235}U **fuel rods** need to be placed in a **moderator** (for example, **water**).

5) If the chain reaction is **left to continue unchecked**, large amounts of **energy** are **released** in a very **short time**, which is very **dangerous**.

6) The **chain reaction** can be controlled by **limiting** the number of **neutrons** in the reactor with **control rods**. These **absorb neutrons** so that the **rate of fission** is controlled.

7) **Coolant** is sent around the reactor to **remove heat** produced in the fission — often the coolant is the **same water** that is being used in the reactor as a **moderator**.

8) The **heat** can then be used to make **steam** for powering **electricity generating turbines**.

Plutonium *is a* Hazardous *Product of* Uranium Fission

 *OCR A Option 04 only*

1) Nuclear reactors produce **radioactive** plutonium as a **waste product**.

2) The **uranium-238** in the fuel rods can **absorb a neutron** to make **plutonium-239**.

3) This **plutonium** is a highly **hazardous** material that can cause a lot of **damage** to the **environment** if it's not handled and **stored carefully** (it is also what is used inside nuclear bombs).

4) One of the biggest **problems** with plutonium is that it undergoes α decay with a very **long half-life** of over **24 000 years** — so it **continues** to be **dangerous** for tens of thousands of years.

Practice Questions

Q1 What is spontaneous fission?

Q2 How can fission be induced in ^{235}U?

Q3 Sketch a diagram of a typical nuclear reactor.
Explain the function of: a) the moderator   and   b) the control rods.

Exam Questions

Q1 (a) Complete this equation for the fission of a ^{235}U nucleus: [2 marks]

$$^{235}_{92}U + {}^{1}_{0}n \rightarrow {}^{90}_{...}Kr + {}^{...}_{56}Ba + 2{}^{1}_{0}n + energy$$

(b) Sketch a graph of fission yield against nucleon number for the fissile products of ^{235}U.   [2 marks]

(c) Mark possible locations for the fissile products Kr and Ba on your graph and label them.   [2 marks]

If anyone asks, I've gone fission...

So, controlled nuclear fission reactions can provide a shedload of energy to generate electricity. There are pros and cons to using fission reactors. They produce huge amounts of energy without emitting any greenhouse gases, but they leave behind some very nasty radioactive waste... But then, you already knew that — now you need to learn all the grisly details.

Nuclear Fusion

This page is for AQA A, AQA B, OCR A and OCR B only.

Nuclear fusion could provide all the energy we need, without the hazardous radioactive waste produced by fission. The trouble is that it's a lot trickier getting nuclei to join than to break them apart — but that doesn't stop those clever types in the white lab coats from giving it a jolly good try...

Fusion Means Joining Nuclei Together

1) **Two light nuclei** can **combine** to create a larger nucleus — this is called **nuclear fusion**.

2) A lot of **energy** is released during nuclear fusion because the new, heavier nuclei have a **much higher binding energy per nucleon** (see p. 90).

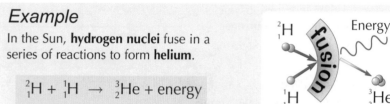

Example
In the Sun, **hydrogen nuclei** fuse in a series of reactions to form **helium**.

$$^2_1\text{H} + ^1_1\text{H} \rightarrow ^3_2\text{He} + \text{energy}$$

Nuclei Need Lots of Energy to Fuse

1) All nuclei are **positively charged** — so there will be an **electrostatic** (or Coulomb) **force** of **repulsion** between them.

2) Nuclei can only **fuse** if they **overcome** this electrostatic force and get **close** enough for the attractive force of the **strong interaction** to hold them both together.

3) About **1 MeV** of kinetic energy is **needed** to make nuclei fuse together — and that's **a lot of energy**.

Low energy nuclei are deflected by electrostatic repulsion

High energy nuclei overcome electrostatic repulsion and are attracted by the strong interaction

Very High Temperatures are Needed for Fusion

1) The **energy** emitted by the **Sun** comes from nuclear **fusion** reactions.

2) **High temperatures** in the Sun mean that nuclei have enough kinetic energy to **fuse together**.

3) **Temperature, T**, can be estimated using the formula: | $E_k = 2 \times 10^{-23} T$ | where E_k is the kinetic energy of the particles in joules.

4) Since **fusion** will only happen if nuclei have a **kinetic energy** of 1 MeV (= 1.6×10^{-13} J), the temperature needed for fusion is in the region of **8×10^9 K**. [This is the temperature at which the <u>average</u> nucleus will have enough energy for fusion. Some nuclei will be much more energetic than the average (see the speed distribution graph on p. 70), so fusion will actually happen at much lower temperatures than this. In prototype fusion reactors the ignition temperature is somewhere around 10^6–10^7 K.]

Plasma is Formed at High Temperatures

1) At the **very high temperatures** needed for nuclear fusion, **atoms don't exist** — the negatively charged electrons are **stripped away**, leaving **positively charged nuclei** and **free electrons**.

2) The resulting mixture is called a **plasma**.

3) Since the **particles** in a plasma are charged, they are affected by **magnetic fields**.

4) So, in **theory** we can use a correctly shaped magnetic field to **confine** the plasma, keeping it **moving** continuously around a **loop**.

Nuclear Fusion

This page is for OCR A Option 04 *only.*

Practical **Fusion Reactors** are **Difficult** to Build

1) Nuclear **fusion** releases **more energy** per nucleon than fission — this makes a **fusion reactor** an attractive option for **generating** the power that we need.

2) An experimental prototype fusion reactor is the **Joint European Torus** (**JET**).

3) Think of JET like a **big rubber ring** that has **magnetic fields** to keep a stream of **plasma** flowing around in a **loop** without touching the sides.

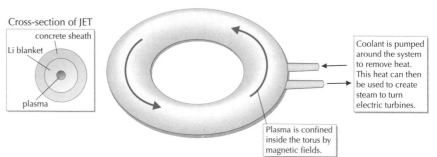

Cross-section of JET
concrete sheath
Li blanket
plasma

Coolant is pumped around the system to remove heat. This heat can then be used to create steam to turn electric turbines.

Plasma is confined inside the torus by magnetic fields.

A torus is just a ring or doughnut shape.

4) JET aims to fuse together **deuterium**, ^{2}H, and **tritium**, ^{3}H, — this reaction is known as **D–T fusion**.

5) The **first thing** that the JET reactor needs to do is **create a plasma** — which is done by passing a whopping **million ampere** electric current through **hydrogen gas** (a mixture of ^{2}H and ^{3}H).

6) The plasma is then **confined** inside the torus by **magnetic fields** (the incredibly high temperature of the plasma would **melt** any material it touched).

7) A **lithium blanket** surrounds the plasma — when this is **bombarded by neutrons** from D–T fusion it reacts to form helium and tritium. The **tritium** can then be fed straight back into the **plasma**.

8) **Coolant** is pumped around the lithium blanket to **remove** the **energy** released during fusion — this **energy** can then be used to create **steam** to turn **electric turbines**.

9) **Unfortunately**, at the moment, the **electricity** generated is **less than** the amount needed to **create the plasma** — so this isn't a **practical design** for commercial use (though it could be if built on a big enough scale).

10) On the plus side, fusion creates no significant **radioactive waste** and the **raw materials** are in **plentiful** supply.

Practice Questions

Q1 Why is a lot of energy required for nuclear fusion to occur?

Q2 Explain what is meant by the term plasma.

Q3 How can a plasma be confined?

Q4 Which particles are caused to fuse in the JET reactor?

Exam Questions

Q1 The kinetic energy required for two protons to fuse is 1.1×10^{-13} J.
Find the temperature required for average nuclei to fuse, given that $E_k = 2 \times 10^{-23}T$. [1 mark]

Q2 The D-T reaction can be written as:
$^2_1$H + $^3_1$H $\rightarrow$ $^4_2$He + $^1_0$n + energy

The following masses should be used when answering these questions:
Deuterium nucleus = 2.013553 u, tritium nucleus = 3.015501 u, helium nucleus = 4.001505 u and neutron = 1.008665 u.

(a) Calculate the total mass defect for this reaction. [2 marks]

(b) How much energy is released in D-T fusion, if a mass defect of 1 u releases 931 MeV of energy? [1 mark]

As the saying goes — 'the fusion is mightier than the fission'...

...well, kinda. Basically, more energy's released per nucleon in a fusion reaction than a fission reaction. But it's hard to get the little blighters to stick together, what with the electrostatic repulsion between nuclei — so fusion reactions only happen at really high temperatures. Possible fusion reactors based on the D-T reaction are being researched at the moment...

Classification of Particles

These pages are for OCR A Option 04 and OCR B only.

There are loads of different types of particle apart from the ones you get in normal matter (protons, neutrons, etc.). They only appear in cosmic rays and in particle accelerators, and they often decay very quickly, so they're difficult to get a handle on. Nonetheless, you need to learn about a load of them and what their properties are.

Don't expect to really understand this (I don't) — but you only need to learn it. Stick with it — you'll get there.

Hadrons are Particles that Feel the Strong Interaction (e.g. Protons and Neutrons)

1) The **nucleus** of an atom is made up from **protons** and **neutrons** held together by the strong interaction (déjà vu).
2) **Not all particles** can **feel** the **strong interaction** — the ones that **can** are called **hadrons**.
3) Hadrons aren't **fundamental** particles. They're made up of **smaller particles** called **quarks** (see pages 100-101).
4) There are **two** types of **hadron** — **baryons** and **mesons**.

Protons and Neutrons are Baryons

1) It's helpful to think of **protons** and **neutrons** as **two versions** of the **same particle** — the **nucleon**. They just have **different electric charges**.
2) As well as **protons** and **neutrons**, there are **other baryons** that you don't get in normal matter — like **sigmas** (Σ) — they're **short-lived** and you **don't** need to **know about them** for A2 (woohoo!).

The Proton is the Only Stable Baryon

All baryons except protons decay to a **proton**. Most physicists think that protons don't **decay**.

Some theories predict that protons should decay with a *very long half-life* of about 10^{32} years — but there's *no evidence* for it at the moment.

Baryon and Meson felt the strong interaction.

The Number of Baryons in a reaction is called the Baryon Number

Baryon number is the number of baryons. (A bit like **nucleon number** but including unusual baryons like Σ too.) The **proton** and the **neutron** each have a baryon number **B = +1**. The **total baryon number** in **any** particle reaction **never changes**.

The Mesons You Need to Know About are Pions and Kaons

1) **All mesons** are **unstable** and have **baryon number B = 0** (because they're not baryons).
2) **Pions** (π-mesons) are the **lightest mesons**. You get **three versions** with different **electric charges** — π^+, π^0 and π^-. Pions were **discovered** in **cosmic rays**. You get **loads** of them in **high energy particle collisions** like those studied at the **CERN** particle accelerator.
3) **Kaons** (K-mesons) are **heavier** and more **unstable** than **pions**. You get different ones like K^+, K^- and K^0.
4) Mesons **interact** with **baryons** via the **strong interaction**.

Pion interactions swap p's with n's and n's with p's, but leave the overall baryon number unchanged.

Summary of Hadron Properties

DON'T PANIC if you don't understand

all this yet. For now, just **learn** these properties. You'll need to work through to the end of page 101 to see how it **all fits in**.

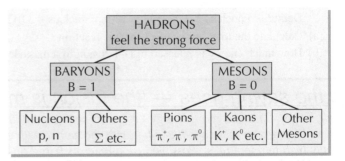

Classification of Particles

Leptons Don't feel the Strong Interaction (e.g. Electrons and Neutrinos)

1) **Leptons** are **fundamental particles** and they **don't** feel the **strong interaction**. The only way they can **interact** with other particles is via the **weak interaction** and gravity (and the electromagnetic force as well if they're charged).

2) **Electrons (e⁻)** are **stable** and very **familiar** but — you guessed it — there are also **two more leptons** called the **muon (μ⁻)** and the **tau (τ⁻)** that are just like **heavy electrons**.

3) **Muons** and **taus** are **unstable**, and **decay** eventually into **ordinary electrons**.

4) The **electron, muon** and **tau** leptons each come with their **own neutrino**: ν_e, ν_μ and ν_τ. *ν is the Greek letter "nu".*

5) **Neutrinos** have **zero** or **almost zero mass** and **zero electric charge** — so they don't do much. **Neutrinos** only take part in **weak interactions** (see p. 101). In fact, a neutrino can **pass right through the Earth** without **anything** happening to it.

You Have to Count the Three Types of Lepton Separately

Each lepton is given a **lepton number** of **+1**, but the **electron, muon** and **tau** types of lepton have to be **counted separately**.

You get **three different** lepton numbers L_e, L_μ and L_τ.

Like the baryon number, the lepton number is just the number of leptons.

Name	Symbol	Charge	L_e	L_μ	L_τ
electron	e^-	−1	+1	0	0
electron neutrino	ν_e	0	+1	0	0
muon	μ^-	−1	0	+1	0
muon neutrino	ν_μ	0	0	+1	0
tau	τ^-	−1	0	0	+1
tau neutrino	ν_τ	0	0	0	+1

Neutrons Decay into Protons

The **neutron** is an **unstable particle** that **decays** into a **proton**. (But it's much more stable when it's part of a nucleus.) It's really just an **example** of β⁻ decay which is caused by the **weak interaction**.

$$n \rightarrow p + e^- + \bar{\nu}_e$$

Free neutrons (i.e. ones not held in a nucleus) have a half-life of about 15 minutes.

The antineutrino has $L_e = -1$ so the total lepton number is zero. Antineutrino? Yes, well I haven't mentioned antiparticles yet. Just wait for the next page …

Practice Questions

Q1 List the differences between a hadron and a lepton.

Q2 Which is the only stable baryon?

Q3 A particle collision at CERN produces 2 protons, 3 pions and 1 neutron. What is the total baryon number of these particles?

Q4 Which two particles have lepton number $L_\tau = +1$?

Exam Questions

Q1 List all the decay products of the neutron. Explain why this decay cannot be due to the strong interaction. [3 marks]

Q2 Initially the muon was incorrectly identified as a meson. Explain why the muon is not a meson. [3 marks]

Go back to the top of page 96 — do not pass GO, do not collect £200...

Do it. Go back and read it again. I promise — read these pages about 3 or 4 times and you'll start to see a pattern. There are hadrons that feel the force, leptons that don't. Hadrons are either baryons or mesons, and they're all weird except for those well-known baryons: protons and neutrons. There are loads of leptons, including good old electrons.

Antiparticles

These pages are for OCR A Option 04 and OCR B only.

More stuff that seems to laugh in the face of common sense — but actually, antiparticles help to explain a lot in particle physics... (Oh, and if you haven't read pages 96 and 97 yet then go back and read them now — no excuses, off you go...)

Antiparticles were Predicted Before they were Discovered

When **Paul Dirac** wrote down an equation obeyed by **electrons**, he found a kind of **mirror image** solution.

1) It predicted the existence of a particle like the **electron** but with **opposite electric charge** — the **positron**.

2) The **positron** turned up later in a cosmic ray experiment. Positrons are **antileptons** so $L_e = -1$ for them. They have **identical mass** to electrons but they carry a **positive** charge.

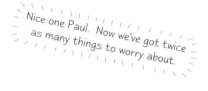

Nice one Paul. Now we've got twice as many things to worry about.

Every Particle has an Antiparticle

Each particle type has a **corresponding antiparticle** with the **same mass** but with **opposite charge**. For instance, an **antiproton** is a **negatively charged** particle with the same mass as the **proton**.

Even the shadowy **neutrino** has an antiparticle version called the **antineutrino** — it doesn't do much either.

Particle	Symbol	Charge	B	L_e	Antiparticle	Symbol	Charge	B	L_e
proton	p	+1	+1	0	antiproton	$\bar{p}$	−1	−1	0
neutron	n	0	+1	0	antineutron	$\bar{n}$	0	−1	0
electron	e	−1	0	+1	positron	e^+	+1	0	−1
electron neutrino	ν_e	0	0	+1	electron antineutrino	$\bar{\nu}_e$	0	0	−1

You can Create Matter and Antimatter from Energy

You've probably heard about the **equivalence** of energy and mass. It all comes out of Einstein's special theory of relativity. **Energy** can turn into **mass** and **mass** can turn into **energy** if you know how — all you need is the formula $E = mc^2$.

It's a good thing this doesn't randomly happen all the time or else you could end up with cute bunny rabbits popping up and exploding unexpectedly all over the place. Oh, the horror...

As you've probably guessed, there's a bit **more to it** than that:

When **energy** is converted into **mass** you have to make **equal amounts** of **matter** and **antimatter**.

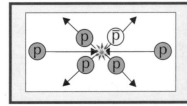

Fire **two protons** at each other at high speed and you'll end up with a lot of **energy** at the point of impact. This energy can be used to make **more particles**.

If you create an extra **proton** you have to create an **antiproton** to go with it. It's called **pair production**.

Antiparticles

Each **Particle-Antiparticle Pair** is Produced from a **Single Photon**

Pair production only happens if **one gamma ray photon** has enough energy to produce that much mass. It also tends to happen near a **nucleus**, which helps conserve momentum.

You usually get **electron-positron** pairs produced (rather than any other pair) — because they have a relatively **low mass**.

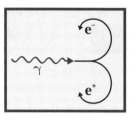

The particle tracks are curved because there's usually a magnetic field present in particle physics experiments (see p. 104). They curve in opposite directions because of the opposite charges on the electron and positron.

The **Opposite** of **Pair Production** is **Annihilation**

When a **particle** meets its **antiparticle** the result is **annihilation**. All the **mass** of the particle and antiparticle gets converted to **energy**. In ordinary matter antiparticles can only exist for a fraction of a second before this happens, so you won't see many of them.

The electron and positron annihilate and their mass is converted into the energy of a pair of gamma ray photons.

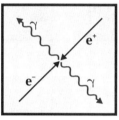

You might wonder why there's any matter at all if antimatter quickly annihilates its partner — but there obviously is, so take a look at p.130 for an explanation.

Mesons are Their **Own Antiparticles** (If you don't know what a meson is, look back at page 96.)

Just before you leave this bit it's worth mentioning that the π^- meson is just the **antiparticle** of the π^+ meson, and the **antiparticle** of a π^0 meson is **itself**. You'll see why on p. 100. So we don't need any more particles here... Phew.

Practice Questions

Q1 Which antiparticle has zero charge and a baryon number of −1?

Q2 Describe the properties of an electron antineutrino.

Q3 What is pair production?

Q4 What happens when a proton collides with an antiproton?

Exam Questions

Q1 Write down an equation for the reaction between a positron and an electron and give the name for this type of reaction. [2 marks]

Q2 According to Einstein, mass and energy are equivalent.
Explain why the mass of a block of iron cannot be converted directly into energy. [2 marks]

Q3 Give a reason why the reaction $p + p \rightarrow p + p + n$ is not possible. [1 mark]

Now stop meson around and do some work...

The idea of every particle having an antiparticle might seem a bit strange, but just make sure you know the main points — a) if energy is converted into a particle, you also get an antiparticle, b) an antiparticle won't last long before it bumps into a particle and annihilates it with a big ba-da-boom, c) this releases the energy it took to make them to start with...

Quarks

These pages are for OCR A Option 04 and OCR B (just "up" and "down" quarks) only.

*If you haven't read pages 96 to 99, do it now! For the rest of you — here are the **juicy bits** you've been waiting for. Particle physics makes **a lot more sense** when you look at quarks. More sense than it did before anyway.*

Quarks are Fundamental Particles

If that first sentence doesn't make much sense to you, <u>read pages 96-99</u> — you have been warned... twice.

Quarks are the **building blocks** for **hadrons** (baryons and mesons).

1) To make **protons** and **neutrons** you only need two types of quark — the **up** quark (**u**) and the **down** quark (**d**).

2) An extra one called the **strange** quark (**s**) lets you make more particles with a property called **strangeness**.

The antiparticles of hadrons are made from **antiquarks**.

Particle physicists have found six different quarks altogether but you only need to know about three of them — up, down and strange (the other three are top, bottom and charm).

Quarks and Antiquarks have Opposite Properties

The **antiquarks** have **opposite properties** to the quarks — as you'd expect.

QUARKS

name	symbol	charge	baryon number	strangeness
up	u	$+\frac{2}{3}$	$+\frac{1}{3}$	0
down	d	$-\frac{1}{3}$	$+\frac{1}{3}$	0
strange	s	$-\frac{1}{3}$	$+\frac{1}{3}$	-1

ANTIQUARKS

name	symbol	charge	baryon number	strangeness
anti-up	$\bar{u}$	$-\frac{2}{3}$	$-\frac{1}{3}$	0
anti-down	$\bar{d}$	$+\frac{1}{3}$	$-\frac{1}{3}$	0
anti-strange	$\bar{s}$	$+\frac{1}{3}$	$-\frac{1}{3}$	$+1$

Baryons are Made from Three Quarks

Evidence for quarks came from **hitting protons** with **high energy electrons** (see page 103).
The way the **electrons scattered** showed that there were **three concentrations of charge** (quarks) **inside** the proton.

Proton = **uud**

Total charge
= 2/3 + 2/3 – 1/3 = 1
Baryon number
= 1/3 + 1/3 + 1/3 = 1

Neutron = **udd**

Total charge
= 2/3 – 1/3 – 1/3 = 0
Baryon number
= 1/3 + 1/3 + 1/3 = 1

Antiprotons are $\bar{u}\bar{u}\bar{d}$ and antineutrons are $\bar{u}\bar{d}\bar{d}$ — so no surprises there then.

Mesons are a Quark and an Antiquark

Pions are just made from **up** and **down** quarks and their **antiquarks**. **Kaons** have **strangeness** so you need to put in **s** quarks as well (remember that the **s** quark has a strangeness of S = –1).

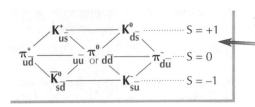

$\underset{u\bar{s}}{K^+}$ ⸻ $\underset{d\bar{s}}{K^0}$ ⸺ S = +1

$\underset{u\bar{d}}{\pi^+}$ ⸺ $\underset{u\bar{u}\ or\ d\bar{d}}{\pi^0}$ ⸺ $\underset{d\bar{u}}{\pi^-}$ ⸺ S = 0

$\underset{s\bar{d}}{\bar{K}^0}$ ⸺ $\underset{s\bar{u}}{K^-}$ ⸺ S = –1

Physicists love patterns. Gaps in patterns like this predicted the existence of particles that were actually found later in experiments. Great stuff.

There's no Such Thing as a Free Quark

What if you **blasted** a **proton** with **enough energy** — could you **separate out** the quarks? Nope. Your energy just gets changed into more **quarks and antiquarks** — it's **pair production** again and you just make **mesons**. This is called **quark confinement**.

Proton — Supply energy to separate the quarks — Quark-antiquark pair produced — Meson

Quarks

The **Weak Interaction** is something that Changes the **Quark Type**

In β⁻ decay a **neutron** is changed into a **proton** — in other words **udd** changes into **uud**.
It means turning a **d** quark into a **u** quark. Only the weak interaction can do this.

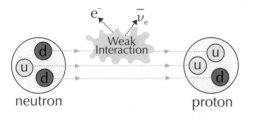

Some unstable isotopes like **carbon-11** decay by β⁺ emission. In this case a **proton** changes to a **neutron**, so a **u** quark changes to a **d** quark and we get:

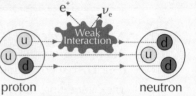

Four Properties are **Conserved** in **Particle Reactions**

Charge and **Baryon Number** are **Always** Conserved

In **any** particle reaction, the **total charge** after the reaction must equal the total charge before the reaction. The same goes for **baryon number**.

Strangeness is Conserved in **Strong Interactions**

The **only** way to change the **type** of quark is with the **weak interaction**, so in strong interactions there has to be the same number of strange quarks at the beginning as at the end. The reaction $K^- + p \rightarrow n + \pi^0$ is fine for **charge** and **baryon number** but not for **strangeness** — so it won't happen. The negative kaon has an s quark in it.

Conservation of **Lepton Number** is a Bit More **Complicated**

The **three types** of lepton number have to be conserved **separately**.
1) For example, the reaction
   $\pi^- \rightarrow \mu^- + \bar{\nu}_\mu$ has $L_\mu = 0$ at the start and $L_\mu = 1 - 1 = 0$ at the end, so it's OK.
2) On the other hand, the reaction $\nu_\mu + \mu^- \rightarrow e^- + \nu_e$ can't happen.
   At the start $L_\mu = 2$ and $L_e = 0$ but at the end $L_\mu = 0$ and $L_e = 2$.

Practice Questions

Q1 What is a quark?
Q2 Which type of particle is made from a quark and an antiquark?
Q3 Describe how a neutron is made up from quarks.
Q4 List four quantities that are conserved in particle reactions.

Exam Questions

Q1 Give the quark composition of the π⁻ and explain how the charges of the quarks give rise to its charge. [2 marks]

Q2 Explain how the quark composition is changed in the β⁻ decay of the neutron. [2 marks]

Q3 Give two reasons why the reaction $p + p \rightarrow p + K^+$ does not happen. [2 marks]

A physical property called strangeness — how cool is that...

True, there's a lot of information here, but this page really does **tie up** a lot of the stuff on the last few pages. Learn as much as you can from this double page, then **go back** to page 96, and **work back** through to here. **Don't expect to** understand it all — but you will **definitely** find it **much easier to learn** when you can see how all the bits **fit in together**.

Particle Accelerators

These pages are for AQA B, OCR A Option 04 and OCR B only.

Particle accelerators are devices that (surprisingly) accelerate particles, using electric and magnetic fields. Accelerated particles can be used to investigate the fundamental particles that make up matter...

Particle Accelerators Cause High-Energy Collisions

1) A **linear accelerator** is a long **straight** tube containing a series of **electrodes**.
2) **Alternating current** is applied to the electrodes so that their **charge** continuously **changes** between + and –.
3) The alternating current is **timed** so that the particles are always **attracted** to the **next electrode** in the accelerator and **repelled** from the **previous** one.
4) A particle's **speed** will **increase** each time it **passes** an electrode — so if the accelerator is long enough particles can be made to approach the **speed of light**.
5) The **high-energy particles** leaving a linear accelerator **collide** with a **fixed target** at the end of the tube.

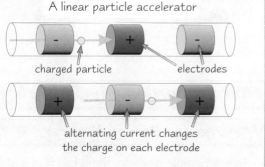
A linear particle accelerator
charged particle   electrodes
alternating current changes the charge on each electrode

6) A **circular accelerator** uses **magnetic fields** to guide the particles around in a circle, accelerating them as they go around (see the **synchrotron** on the next page).
7) Often **two beams** are accelerated in **opposite directions** and caused to crash into each other at specific **detection points**.
8) **Colliding beam accelerators** produce impacts with a **greater energy** than fixed target accelerators — so they can probe **deeper** into matter. However, it is **easier to detect** the effect of impacts in **fixed target** accelerators.

A Cyclotron can Accelerate Protons up to 20 MeV

1) A cyclotron uses **two semicircular electrodes** to accelerate protons or other charged particles across a gap.
2) An **alternating potential difference** is applied between the electrodes — as the **particles** are **attracted** from one side to the other their **energy increases** (i.e. they are **accelerated**).
3) A **magnetic field** is used to keep the particles moving in a **circular motion** (in the diagram on the right, the magnetic field would be perpendicular to the page).
4) The combination of the **electric** and **magnetic fields** makes the particles **spiral outwards** as their energy increases.
5) This will only happen if the **supply frequency**, *f*, of the alternating potential difference is at a **certain value**.

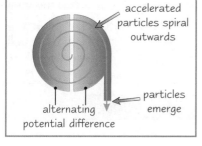
accelerated particles spiral outwards
alternating potential difference
particles emerge

Example   Derive the supply frequency needed to accelerate a particle inside a cyclotron.

1) The force on a charged particle moving in a magnetic field, $F = Bqv$
2) The particles move in a circle, so F gives the centripetal force $\Rightarrow Bqv = \dfrac{mv^2}{r}$
3) Rearranging for v gives: $v = \dfrac{Bqr}{m}$
4) The time of orbit $T = \dfrac{\text{distance}}{\text{speed}} = \dfrac{2\pi r \cdot m}{Bqr} = \dfrac{2\pi m}{Bq}$ (where $2\pi r$ is the circumference of a circle)
5) So, supply frequency, $f = \dfrac{1}{T} = \dfrac{Bq}{2\pi m}$

If particles are accelerated close to the **speed of light**, their **mass** increases, which means they take **longer** to travel around the cyclotron. This effect **limits** the **energy** available in a cyclotron to around **20 MeV** for protons.

Particle Accelerators

Synchrotrons Produce Very High Energy Beams

1) A **synchrotron** is a very large **circular particle accelerator** that is able to produce particle collisions with much **higher energies** than either a linear accelerator or a cyclotron.

2) This very high energy beam means you can **probe even further** into matter to investigate the properties of **fundamental particles**, such as quarks (see p. 100).

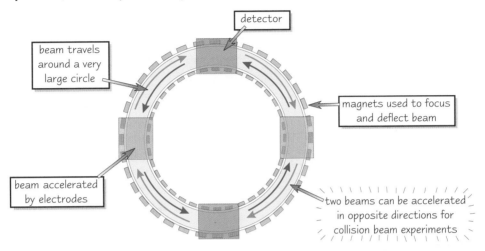

detector

beam travels around a very large circle

magnets used to focus and deflect beam

beam accelerated by electrodes

two beams can be accelerated in opposite directions for collision beam experiments

3) A beam of **charged particles** is accelerated by pairs of **electrodes**, in much the same way as in a linear accelerator.

4) **Two beams** of oppositely charged particles can be accelerated in **different directions** around the synchrotron at the same time — for **collision beam** experiments.

5) A ring of **electromagnets** keeps the particles moving in a **circular path** — this is achieved by increasing the **magnetic field** strength to compensate for increases in **relativistic mass**.

6) Magnets are also used to keep the particles in **focused beams**.

7) In this way, **synchrotrons** can produce much **higher energy** particle beams than **cyclotrons** — with particles reaching energies of over **500 GeV**.

8) A **synchrotron** produces **small bursts** of particles, so, if your experiment requires a **continuous stream** of particles, you'll need to use a **cyclotron** instead.

Practice Questions

Q1 Explain how particles are accelerated in a linear accelerator.

Q2 What forces act on a particle in a cyclotron?

Q3 Which type of accelerator is able to produce the highest energy particles?

Q4 Explain how particles are accelerated in a synchrotron.

Exam Questions

Q1 A cyclotron is used to accelerate protons. Calculate the frequency the cyclotron should operate with if the field strength is 20 mT, the charge on a proton is $+1.6 \times 10^{-19}$ C and the mass of a proton is 1.67×10^{-27} kg. [2 marks]

Q2 (a) Explain why there is a limit to the energy of particles accelerated in a cyclotron. [3 marks]

(b) How does the synchrotron overcome the problem identified in part (a) of this question? [2 marks]

(c) State an advantage and a disadvantage of a synchrotron particle accelerator. [2 marks]

Smash high-energy particles together to see what they're made of...

So, the three types of particle accelerator all have their advantages, but the synchrotron wins hands down on making very high energy particles for investigating fundamental particles. A famous synchrotron (in the physics world) is the Large Electron Positron (LEP) collider found at CERN on the Swiss-French border — this accelerator is a whopping 27 km loop...

Detecting Particles

These pages are for AQA B only.

Luckily for us, charged particles affect atoms as they pass by — which means we can see what's going on...

Charged Particles Leave Tracks

When a charged particle passes through a substance it causes **ionisation** — electrons are knocked out of atoms.
The particle leaves a **trail of ions** as it goes.

The easiest way to **detect** the particle is if you somehow make the **trail of ions show up** and then take a **photo**.

Cloud Chambers Work by Supercooling a Gas

You don't need to remember the details of how these chambers work.

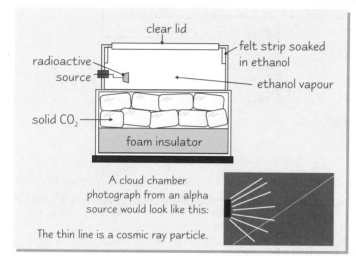

A cloud chamber photograph from an alpha source would look like this:

The thin line is a cosmic ray particle.

Cloud chambers work using a **supercooled vapour** — that's something that's still a gas below its usual condensation temperature. The ions left by particles make the vapour **condense** and you get "**vapour trails**" (a bit like the ones left by jet planes).

Only the **charged particles** show up.

A **magnetic field** makes the particles follow **curved tracks** — the larger the curve radius, the greater the particle's **momentum**. Positive and negative particles curve **opposite** ways — you can find out which is which using **Fleming's** left-hand rule.

Cloud chambers are usually used to look at the products from **radioactive decay**:
Heavy, **short** tracks mean lots of ionisation, so those will be the α-**particles**. Fainter, **long** tracks are β-**particles**.

Bubble Chambers use a Superheated Liquid

Bubble chambers are a bit like cloud chambers in reverse. Hydrogen is kept as a **liquid** above its **boiling point** by putting it under **pressure**. If the pressure is suddenly **reduced**, **bubbles of gas** will start to form in the places where there is a trail of ions. You have to take the photo **quickly** before the bubbles grow too big.

Using **hydrogen** in a bubble chamber means you see collisions with **stationary proton targets** — hydrogen nuclei.

Neutral Particles Only Show Up When They Decay

Remember that **neutral** particles **don't** make tracks.

You can only see them when they **decay**. If you see a **V** shape starting in the middle of nowhere, it will be two oppositely charged particles from the decay of a neutral particle.

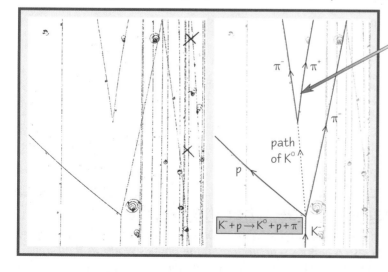

This V comes from the decay $K^0 \rightarrow \pi^+ + \pi^-$

The **distance** from the **interaction point** to the V depends on the **half-life** of the neutral particle. Longer-lived particles travel **further** on average before they decay — but you have to be careful.

The particles are travelling **close to the speed of light** so relativistic **time dilation** (aaarghhh — see p. 136) makes them survive for much longer than normal.

Here the particles have so much momentum that the tracks are almost straight.

Detecting Particles

Real Bubble Chamber Photographs can be a bit Intimidating

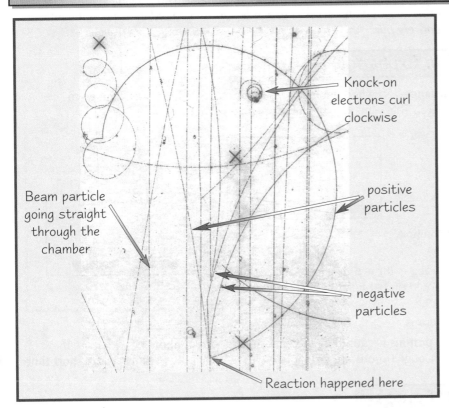

Knock-on electrons curl clockwise

Beam particle going straight through the chamber

positive particles

negative particles

Reaction happened here

At first sight the photo might look a bit of a mess with tracks everywhere. Don't panic — start by finding the incoming beam.

The **straight** lines are from the incoming beam. Several particles will go straight through without doing anything — you can just ignore them.

Look for a little spiral coming from one of the straight tracks. It shows a **knock-on electron** — an electron that's been kicked out of one of the hydrogen atoms. Knock-on electrons tell you **two** things — **which way** the particles are going and which way negative particles **curve**.

Here the particles are going **up** and **negative** ones curl **clockwise**.

Find a **point** with **several** curved tracks coming from it — that's a reaction. You can identify positively and negatively charged particles from the **way they curve**.

Cloud Chambers and Bubble Chambers aren't used Any More

Nowadays, particle physicists use detectors that give out **electrical signals** that are sent **straight** to a **computer**. It's a bit easier than having a whole team of scientists squinting over thousands of photos. Modern detectors include **drift chambers**, **scintillation counters** and **solid state detectors**. You don't need to know any details about these for the exam.

Practice Questions

Q1 Describe how a cloud chamber works.
Q2 Explain the operation of a bubble chamber.
Q3 Which particles don't show up in bubble chamber photos?
Q4 How does the track of an electron show that the electron is losing energy?

Exam Questions

Q1 Explain how the charges of particles can be found from their tracks in bubble chamber photographs. [3 marks]

Q2 The reaction $p + p \rightarrow p + n + \pi^+ + \pi^0$ occurs in a bubble chamber. Which products of this reaction will form tracks? [1 mark]

Q3 A photon, travelling through a bubble chamber, is converted into an $e^- e^+$ pair. Draw a sketch showing the tracks that would be formed by this reaction. [3 marks]

Detecting particles is a piece of cake — with a computer...

*Typical. They now have an easy way of detecting particles — but you have to learn the methods that are a) harder, and b) now completely obsolete. *sigh* I suppose it's quite nice to actually see the pictures though.*

SECTION NINE — NUCLEAR & PARTICLE PHYSICS

Exchange Particles and Feynman Diagrams

These pages are for OCR B only.

*Having learnt about hadrons (baryons and mesons) and leptons, antiparticles and quarks, you now have the esteemed privilege of learning about yet another weirdy thing called a **gauge boson** (where did they get that name?). To the casual observer this might not seem **entirely fair**. And I have to say, I'd be with them.*

Forces are Caused by **Particle Exchange**

You can't have **instantaneous action at a distance**. So when two particles **interact**, something must **happen** to let one particle know that the other one's there. That's the idea behind **exchange particles**.

1) **Repulsion** — Each time the **ball** is **thrown or caught** the people get **pushed apart**. It happens because the ball carries **momentum**.

Particle exchange also explains **attraction** but you need a bit more imagination.

←——REPULSION——→

The particles don't actually loop round like that, though.

——→ATTRACTION←——

2) **Attraction** — Each time the **boomerang** is **thrown or caught** the people **get pushed together**. (In real life, you'd probably fall in first.)

These exchange particles are called **gauge bosons**.

The **electromagnetic repulsion** between two **protons** is caused by the **exchange** of **virtual photons**, which are the gauge bosons of the **electromagnetic** force. Gauge bosons are **virtual** particles — they only exist for a **very short time**.

There are **Four Fundamental Forces**

All forces in nature are caused by these four **fundamental** forces.
Each one has its **own gauge boson** and you have to learn their names:

Type of Interaction	Gauge Boson	Particles Affected
strong	gluon	hadrons only
electromagnetic	photon (symbol, γ)	charged particles only
weak	W^+, W^-, Z^0	all types
gravity	graviton?	all types

Particle physicists never bother about gravity because it's so incredibly feeble compared with the other types of interaction. Gravity only really matters when you've got big masses like stars and planets.

The graviton may exist but there's no evidence for it.

The **Larger** the **Mass** of the **Gauge Boson**, the **Shorter** the **Range** of the **Force**

1) The **W bosons** have a **mass** of about **100 times that of a proton**, so that gives the weak force a **very short range**. Creating a virtual W particle uses **so much energy** that they can only exist for a **very short time** and they **can't travel far**.

2) On the other hand, the **photon** has **zero mass**, so that gives you a force with **infinite range**.

Feynman Diagrams Show What's **Going in** and What's **Coming Out**

Richard Feynman was a brilliant physicist who was famous for explaining complicated ideas in a fun way that actually made sense. He worked out a really **neat way** of **solving problems** by **drawing pictures** rather than doing **calculations**.

1) **Gauge bosons** are represented by **wiggly lines** (technical term).

2) All other **particles** are represented by **straight lines**.

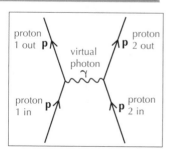

Exchange Particles and Feynman Diagrams

*You can draw Feynman diagrams for **loads** of reactions: here are some examples using the weak interaction.*

Beta-plus and Beta-minus Decay

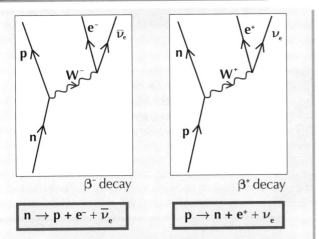

β⁻ decay

β⁺ decay

$$n \rightarrow p + e^- + \bar{\nu}_e$$

$$p \rightarrow n + e^+ + \nu_e$$

You get an **antineutrino** in β⁻ decay and a **neutrino** in β⁺ decay so that **lepton number** is conserved.

RULES FOR DRAWING FEYNMAN DIAGRAMS:

1) **Incoming** particles start at the bottom of the diagram and outgoing particles leave at the top.

2) The **baryons** stay on one side of the diagram, and the **leptons** stay on the other side.

3) The **W** bosons carry **charge** from one side of the diagram to the other — make sure charges balance.

4) A **W⁻** particle going to the **left** has the same effect as a **W⁺** particle going to the **right**.

A Proton Capturing an Electron

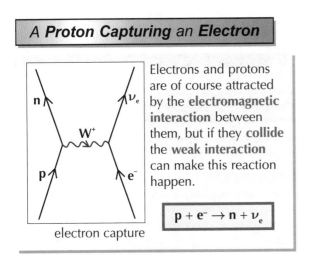

electron capture

Electrons and protons are of course attracted by the **electromagnetic interaction** between them, but if they **collide** the **weak interaction** can make this reaction happen.

$$p + e^- \rightarrow n + \nu_e$$

Neutrinos Interacting with Matter

There's a very **low probability** of a **neutrino interacting with matter**, but here's what happens when one does.

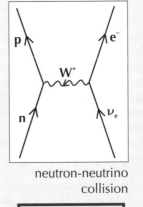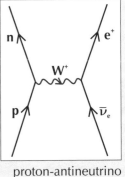

neutron-neutrino collision

proton-antineutrino collision

$$n + \nu_e \rightarrow p + e^-$$

$$p + \bar{\nu}_e \rightarrow n + e^+$$

Practice Questions

Q1 List the four fundamental forces in nature.

Q2 Explain what a virtual particle is.

Q3 Draw a Feynman diagram for a neutrino-neutron interaction.

Q4 Which gauge bosons are exchanged in weak interactions?

Exam Questions

Q1 How is the force of electromagnetic repulsion between two protons explained by particle exchange? [2 marks]

Q2 Draw a Feynman diagram for the collision between an electron antineutrino and a proton. Label the particles and state clearly which type of interaction is involved. [3 marks]

Feynman diagrams are the best thing in particle physics...

These little beauties are great — they are a nice simple way to represent what's going on when particles interact with each other. You don't need to learn the theory behind the diagrams — just learn the ones on these pages and you'll be fine...

The Solar System & Astronomical Distances

This page is for AQA A Option 5 (you don't need the stuff on the Solar System), OCR A Option 01, OCR B and Edexcel (just the definition of the light-year).

Just remember that you're standing on a planet that's evolving, and revolving at 900 miles an hour...

Our **Solar System** Contains the **Sun**, **Planets**, **Satellites**, **Asteroids** and **Comets**

1)   Our **Solar System** consists of the **Sun** and all of the objects that **orbit** it:

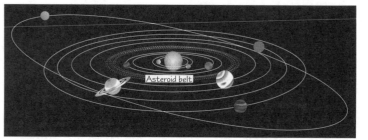

The Planets (in order): **Mercury, Venus, Earth, Mars, Jupiter, Saturn, Uranus, Neptune** and **Pluto** (as well as the Asteroid belt) all have nearly **circular** orbits. Pluto's orbit is the least circular of the planets.

Remember — planets, moons and comets don't emit light; they just reflect it.

2)   The orbits of the **comets** we see are **highly elliptical**. Comets are "**dirty snowballs**" that we think usually orbit the sun about **1000 times further away** than **Pluto** does (in a ring called the Oort cloud). Occasionally one gets **dislodged** and heads towards the Sun. It follows a new elliptical orbit, which can take **millions of years** to complete. Some comets (from closer in than the Oort cloud) follow a **smaller orbit** and they return to swing round the Sun more regularly. The most famous is **Halley's comet**, which orbits in **76 years**.

For most of its life a comet is just a frozen lump. A tail forms on the comet as it comes close enough for the Sun to start to melt it.

The Oort cloud

The tail always points away from the Sun as it is "blown" by the solar wind

Distances and **Velocities** in the Solar System can be Measured using **Radar**

1)   **Radio telescopes** (see p. 112) can be used to send **short pulses** of **radio waves** towards a planet or meteor (a rock flying about the solar system), which **reflect** off the surface and bounce back.

2)   The telescope picks up the reflected radio waves and the **time taken** (*t*) for them to return is measured.

3)   Since we know the **speed** of radio waves (**speed of light, *c***) we can work out the **distance**, *d*, to the object using:

$$2d = ct$$

It's 2d, not just d, because the pulse travels twice the distance to the object — there and back again.

4)   If **two** short pulses are sent a certain **time interval** apart, you can measure the **distance** an object has moved in that time. From this time and distance, you can calculate the **average speed** of the object **relative** to Earth. More accurate measurements can be made using the Doppler shift (see p. 128).

Distances in the Solar System are Often Measured in **Astronomical Units (AU)**

1)   From **Copernicus** onwards, astronomers were able to work out the **distance** the **planets** are from the Sun **relative** to the Earth, using **astronomical units** (AU). But they could not work out the **actual distances**.

One **astronomical unit (AU)** is defined as the **mean distance** between the **Earth** and the **Sun**.

2)   The **size** of the AU wasn't known until 1769 — when it was carefully **measured** during a **transit of Venus**.

Another Measure of Distance is the **Light-Year (ly)**

1)   All **electromagnetic waves** travel at the **speed of light**, *c*, in a vacuum ($c = 3.00 \times 10^8$ ms^{-1}).

The **distance** that electromagnetic waves travel in **one year** is called a **light-year (ly)**.

2)   If we see the light from a star that is, say, **10 light-years away** then we are actually seeing it as it was **10 years ago**. The further away the object is, the further **back in time** we are actually seeing it.

3)   **1 ly** is equivalent to about **63 000 AU**.

The Solar System & Astronomical Distances

This page is for AQA A Option 5, OCR A Option 01 and OCR B (not the parallax bit) only.

The Distance to **Nearby Stars** can be Measured by **Parallax**

1) You experience parallax every day.
Imagine you're in a **moving car**. You see that (stationary) objects in the **foreground** seem to be **moving faster** than objects in the **distance**.

2) This **apparent change in position** is called **parallax** and is measured in terms of the **angle of parallax**. The **greater** the **angle**, the **nearer** the object is to you.

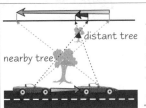

The nearby tree seems to have moved much further relative to the horizon than the more distant tree. The angles marked are called <u>angles of parallax</u>.

3) The distance to **nearby stars** can be calculated by observing how they **move relative** to **very distant stars** when the Earth is in **different parts** of its **orbit**. This gives a **unit** of distance called a **parsec (pc)**.

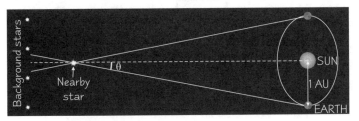

A star is exactly **one parsec (pc)** away from Earth if the **angle of parallax**,

$$\theta = 1 \text{ arcsecond} = \left(\frac{1}{3600}\right)^{\circ}$$

Important **Sizes** and **Conversions**

Unit of Distance	Astronomical Unit (AU)	Light Year (ly)	Parsec (pc)
Approximate Length in metres	1.50×10^{11}	9.46×10^{15}	3.09×10^{16}

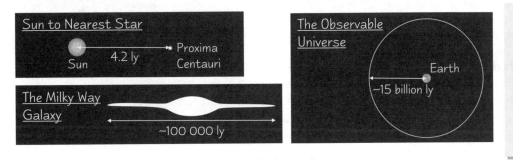

<u>Sun to Nearest Star</u> — Sun — 4.2 ly → Proxima Centauri

<u>The Milky Way Galaxy</u> — ~100 000 ly

<u>The Observable Universe</u> — Earth — ~15 billion ly

When we look at the stars we're looking **back in time**, and we can only see as far back as the **beginning of the Universe**.

So the **size** of the **observable Universe** is the **age** of the Universe multiplied by the **speed of light**.

Practice Questions

Q1 What are the principal contents of our Solar System?
Q2 What is meant by a) an astronomical unit, b) a parsec and c) a light-year?
Q3 How do we measure the distance to the objects in the Solar System using radar?

Exam Questions

Q1 (a) Outline the main differences between planets and comets. [5 marks]
(b) Explain why a comet has a tail which always points away from the Sun. [2 marks]

Q2 (a) Give the definition of a *light-year*. [1 mark]
(b) Calculate the distance of a light-year in metres. [2 marks]
(c) Why is the size of the observable universe limited by the speed of light? [2 marks]

So — using a ruler's out of the question then...

Don't bother trying to get your head round these distances — they're just too big to imagine. Just learn the powers of ten and you'll be fine. Make sure you understand the definition of a parsec — it's a bit of a weird one.

Telescopes and Detectors

These pages are for AQA A Option 5 only.

Some optical telescopes use lenses (no, really), so first, here's a bit of lens theory...

Converging Lenses Bring Light Rays Together

1) **Lenses** change the **direction** of light rays by **refraction**.

2) Rays **parallel** to the **principal axis** of the lens converge onto a point called the **principal focus**. Parallel rays that **aren't** parallel to the principal axis converge somewhere else on the **focal plane** (see refracting telescope).

3) The **focal length**, *f*, is the distance between the **lens axis** and the **principal focus**.

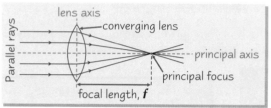

4) A **more powerful** (thicker) lens converges the rays more **strongly** and will have a **short focal length**.

5) The power of a lens with focal length *f* m is:

$$P = \frac{1}{f}$$

where lens power is measured in **dioptres**, **D**.

Images can be Real or Virtual

1) A **real image** is formed when light rays from an object are made to **pass through** another point in space. The light rays are **actually there**, and the image can be **captured** on a **screen**.

2) A **virtual image** is formed when light rays from an object **appear** to have come from another point in space. The light rays **aren't really there**, so the image **can't** be captured on a screen.

3) Converging lenses can form both **real** and **virtual** images, depending on where the object is. If the object is **further** than the **focal length** away from the lens, the image is **real**. If the object's **closer**, the image is **virtual**.

4) To work out where an image will appear, you can draw a **ray diagram**. You only need to draw **two rays** on a ray diagram: one **parallel** to the principal axis that passes through the **principal focus**, and one passing through the **centre** of the lens that **doesn't get bent**.

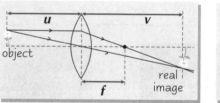

In the diagram, *u* = distance between object and lens axis, *v* = distance between image and lens axis (**positive** if image is **real**, **negative** if image is **virtual**), *f* = focal length.

5) The values *u*, *v* and *f* are related by the **lens equation**: $\Longrightarrow$ $\frac{1}{f} = \frac{1}{u} + \frac{1}{v}$

A Refracting Telescope uses Two Converging Lenses

1) The **objective lens** converges the rays from the object to form a **real image**.

2) The **eye lens** acts as a **magnifying glass** on this real image to form a **magnified virtual image**.

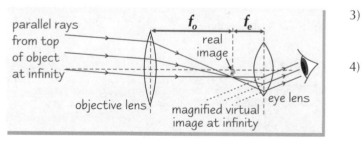

3) If you assume the object is at infinity, then the rays from it are **parallel**, and the real image is formed on the **focal plane**.

4) A **telescope** (in normal adjustment) is set up so that the **principal focus** of the **objective** lens is in the **same position** as the principal focus of the **eye** lens, so the **final magnified image** appears to be at **infinity**.

5) The **magnification**, *M*, of the telescope can be calculated in terms of angles, or the focal length. The **angular magnification** is the **angle** subtended by the **image** θ_i over the **angle** subtended by the **object** θ_o at the eye:

$$M = \frac{\theta_i}{\theta_o}$$

or in terms of **focal length** (with the telescope in normal adjustment as shown above):

$$M = \frac{f_o}{f_e}$$

Telescopes and Detectors

A *Reflecting Telescope* uses a *Concave Mirror* and a *Converging Lens*

1) A **concave mirror** (called the **objective mirror**) converges parallel rays from an object, forming a **real image**.

2) An **eye lens magnifies** the image as before.

3) The focal point of the mirror (where the image is formed) is **in front** of the mirror and so an arrangement needs to be devised so that the observer doesn't **block out** the light. A set-up called the **Cassegrain arrangement** is a common solution to this problem: ⟹

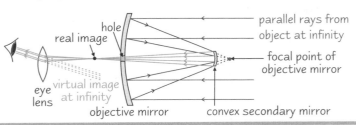

The *Resolving Power* of a Telescope — how much *Detail* you can See

1) The **resolving power** of an instrument is the **smallest angle** at which it can **distinguish** two points.

Two stars that can only just be distinguished

θ = resolving power

About half of the stars that we see in the night sky are actually collections of two or more stars. Our eyes see them as a single star since the angle between them is too small to resolve.

2) Resolution is limited by diffraction. If a beam of light passes through a circular **aperture**, then a **diffraction pattern** is formed. The central circle is called the **Airy disc** (see p.64 for an example of the pattern).

3) **Two** light sources can **just** be distinguished if the **centre** of the **Airy disc** from one source is **at least as far away** as the **first minimum** of the other source. This led to the Rayleigh criterion:

$$\theta \approx \frac{\lambda}{D}$$

where θ is the **minimum angle** that can be resolved in **radians**, λ is the **wavelength** of the light in **metres** and D is the **diameter** of the **aperture** in **metres**.

4) For **telescopes**, D is the diameter of the **objective lens** or the **objective mirror**. So very large lenses or mirrors are needed to see **fine detail**.

There are *Big Problems* with *Refracting Telescopes*

1) Glass refracts **different colours** of light by **different amounts** and so the image for each colour is in a slightly **different position**. This **blurs** the image and is called **chromatic aberration**.

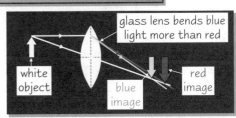

2) Any **bubbles** and **impurities** in the glass **absorb** some of the light, which means that **very faint** objects **aren't seen**. Building large lenses that are of a **sufficiently good quality** is **difficult** and **expensive**.

3) **Large lenses** are very **heavy** and can only be **supported** from their **edges**, so their **shape** can become **distorted**.

4) For a **large magnification**, the **objective lens** needs to have a **very long focal length**. This means that refracting telescopes have to be **very long**, leading to very **large** and **expensive buildings** needed to house them.

Reflecting telescopes are better, but not perfect — see next page →

Practice Questions

Q1 Define the focal length and the power of a converging lens.

Q2 Draw ray diagrams to show how an image is formed in a refracting and a reflecting (Cassegrain) telescope.

Q3 Explain resolving power and the Rayleigh criterion.

Complicated things, telescopes — and there's more on the next page...

Refracting telescopes are sometimes set up with the focal points of the lenses in different positions. If they're set up just right, the final image is real and you can project it on a screen. So you can look at the Sun without going blind.

Telescopes and Detectors

These pages are for AQA A Option 5 only.

Previously on Telescopes and Detectors... big refracting telescopes are a bit dodgy...

Reflecting Telescopes are *Better* than Refractors but they have *Problems* too

1) **Large mirrors** of **good quality** are much **cheaper** to build than large lenses. They can also be **supported** from **underneath** so they don't **distort** as much as lenses.

2) Mirrors don't suffer from **chromatic aberration** (see p. 111) but can have **spherical aberration**:

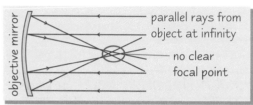

parallel rays from object at infinity

no clear focal point

objective mirror

If the **shape** of the mirror isn't quite **parabolic**, parallel rays reflecting off different parts of the mirror do not all **converge** onto the same point.

When the **Hubble Space Telescope** was first launched it suffered from **spherical aberration**. They had to find a way round the problem before it could be used.

Radio Telescopes are *Similar* to *Optical* Telescopes in Some Ways

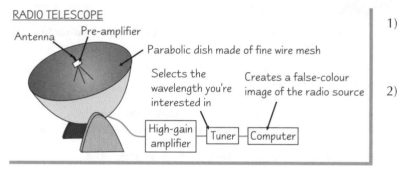

RADIO TELESCOPE

Antenna   Pre-amplifier

Parabolic dish made of fine wire mesh

Selects the wavelength you're interested in

Creates a false-colour image of the radio source

High-gain amplifier — Tuner — Computer

1) The most obvious feature of a radio telescope is its **parabolic dish**. This works in exactly the same way as the **objective mirror** of an **optical reflecting** telescope.

2) An **antenna** is used as a detector at the **focal point** instead of an eye or camera in an optical telescope, but there is **no equivalent** to the **eye lens**.

3) Most radio telescopes are **manoeuvrable**, allowing the source of the waves to be **tracked** (in the same way as optical telescopes). The telescope moves with the source, stopping it 'slipping out of view' as the Earth rotates.

The *Bigger* the *Dish* the *Greater* the *Gathering Power*

1) A **bigger dish** (or mirror for optical reflectors) collects **more energy** from the object that they're looking at in a given time. This gives a **more intense image**, allowing the telescope to observe **fainter** objects.

2) The **gathering power** (energy gathered per second) is proportional to the area:

$$Power \propto Diameter^2$$

The bigger the dish, the greater the gathering power. Mmm....

Radio Waves have a Much *Longer Wavelength* than Light

1) The **wavelengths** of **radio waves** are about a **million times longer** than the wavelengths of **light**.

2) The **Rayleigh criterion** (see p. 111) gives the **resolving power** of a telescope as $\theta \approx \lambda/D$.

3) So for a radio telescope to have the **same resolving power** as an optical telescope, its dish would need to be a **million times bigger** (about the size of the UK for a decent one). The **resolving power** of a radio telescope is **worse** than the **unaided eye**.

Radio astronomers get around this by linking lots of telescopes together. Using some nifty computer programming, their data can be combined to form a single image. This is equivalent to one huge dish the size of the separation of the telescopes. Resolutions thousands of times better than optical telescopes can be achieved this way.

4) Instead of a **polished mirror**, a **wire mesh** can be used since the long wavelength radio waves don't notice the gaps. This makes their **construction** much **easier** and **cheaper** than optical reflectors.

5) The **shape** of the dish has to have a **precision** of about $\lambda/20$ to avoid **spherical aberration** (see above). So the dish does not have to be **anywhere near as perfect** as a mirror.

6) Unlike an optical telescope, a radio telescope has to **scan across** the radio source to **build up** the **image**.

Telescopes and Detectors

Charge-Coupled Devices (CCDs) are Very Sensitive Light Detectors

1) CCDs are **silicon chips** about the size of a postage stamp, divided up into a grid of millions of **identical pixels**.

2) Silicon is a **semiconductor** so it doesn't usually have many **free electrons**. When **light** shines on a pixel, **electrons** are released from the silicon, with the **number** of electrons depending on the **brightness** of the light.

3) **Underneath** each **pixel** is a **potential well** (a kind of controllable electrical bucket), which traps the electrons.

4) Once a picture has been taken, the electrons are **shunted** from **one potential well** to **another** so that they all come out **in sequence** from **one corner** of the CCD. (This is called an 'electrical bucket brigade' to use the lingo.)

5) This sequence can be converted into a **digital signal** and sent to computers **anywhere in the world**.

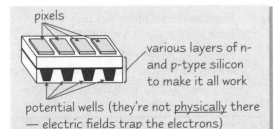

pixels

various layers of n- and p-type silicon to make it all work

potential wells (they're not <u>physically</u> there — electric fields trap the electrons)

CCDs use Quantum Effects

1) **Quantum physics** tells us that EM radiation is formed in **discrete packets** of energy called **photons**.

2) The incoming photons release electrons in the silicon due to the **photoelectric effect**. (See p. 62.)

3) Electrons are released by **more than 70%** of the photons that hit a pixel, so the **quantum efficiency** of a CCD is greater than 70%. On average, the **eye** needs about **100 photons** before it responds and so has a quantum efficiency of about **1%**. The quantum efficiency of a **photographic emulsion** is about **4%**.

Practice Questions

Q1 Why do radio telescopes struggle with their resolving power?

Q2 Why is it easier to make the parabolic dish on a radio telescope than it is to make a concave mirror?

Q3 What does CCD stand for?

Q4 What does quantum efficiency mean and what is the quantum efficiency of a CCD?

Exam Questions

Q1 (a) Define the *principal focus* and the *focal length* of a converging lens. [2 marks]

(b) An object was placed 0.20 m in front of a converging lens of focal length 0.15 m. How far behind the lens was the image formed? [3 marks]

(c) The object was placed 0.10 m in front of the same lens. Where was the image formed? [2 marks]

Q2 An objective lens of focal length 5.0 m and an eye lens of focal length 0.10 m are used in a refracting telescope.

(a) How far apart should the lenses be placed for the telescope to be in normal adjustment? [1 mark]

(b) Define angular magnification and calculate the angular magnification of this telescope. [2 marks]

Q3 (a) How is the gathering power of a telescope related to its objective diameter? [1 mark]

(b) The Arecibo radio telescope dish in Puerto Rico has a diameter of 300 m and the Jodrell Bank dish near Manchester has a diameter of 76 m. Calculate the ratio of their gathering powers. [2 marks]

Q4 (a) Outline the basic function of a CCD. [3 marks]

(b) Why are CCDs much better at taking pictures of very faint objects than conventional film cameras? [3 marks]

CCDs were a quantum leap for astronomy — get it... quantum leap... *sigh*

With CCDs, you can get all the images you want from the comfort of your nearest Internet café. Gone are the days of standing on a hill with a telescope and a thermos hoping the sky clears before your nose turns black and falls off. Shame.

Spectra

This page is for Edexcel, AQA A Option 5 and OCR A Option 01 only.

Oooooo... pretty colours. Useful too apparently...

Spectra are Produced by Electrons Moving Between Energy Levels

1)   **Electrons** in an **atom** can **only exist** at certain **well-defined energy levels**.
     Each level is given a **number**, with *n* = **1** representing the **ground state** (or **lowest energy** state).

2)   Electrons can **move down** an energy level by **emitting** a photon. Since these **transitions** are between **definite energy levels**, the **photons** emitted can **only** take **certain allowed values**.

     (See page 66 for more detail on this.)

Hot Gases Produce Line Emission Spectra

1)   If you heat a **gas** up to a **high temperature**, many of its electrons become **excited** — they move up to **higher energy levels** in their atoms.

2)   As they fall back down to the **ground state**, these electrons emit energy in the form of **photons**.

3)   If you **split** the light from a hot gas with a **prism** or **diffraction grating** (see p. 32), you get a **line spectrum**. A line spectrum is seen as a **series** of **bright lines** against a **black background**.

4)   Each **line** corresponds to a **particular wavelength** of light **emitted** by the source. Since only **certain photon energies** are **allowed** for each element, you can use the spectrum of gas to tell what elements are in it.

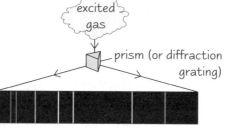

Continuous Spectra Contain All Possible Wavelengths

1)   The **spectrum** of **white light** is **continuous**.

2)   If you **split** the **light** up with a **prism**, the **colours** all **merge** into each other — there **aren't** any **gaps** in the spectrum.

3)   **Hot things** emit a **continuous spectrum** in the visible and infrared (**black body radiation**, see p.118).

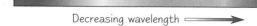

Decreasing wavelength ⟹

Shining White Light through a Cool Gas gives an Absorption Spectrum

1)   You get a **line absorption spectrum** when **light** with a **continuous spectrum** of **energy** passes through a cool gas.

2)   At **low temperatures**, **most** of the **electrons** in the **gas atoms** will be in their **ground states**.

3)   **Photons** of the **correct wavelength** are **absorbed** by the **electrons** to **excite** them to **higher energy levels**.

4)   These **wavelengths** are then **missing** from the **continuous spectrum** when it **comes out** the other side of the gas.

5)   You see a **continuous spectrum** with **black lines** in it corresponding to the **absorbed wavelengths**.

6)   If you **compare** the **absorption** and **emission** spectra of a particular gas, the **black lines** in the **absorption spectrum** **match up** to the **bright lines** in the **emission spectrum**.

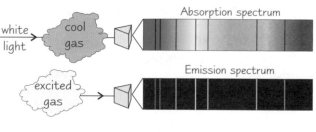

Excited Molecules Emit Band Spectra

1)   You can **excite** a **molecule** by making the **electrons** in its **bonds vibrate**.

2)   The **energies** of these **vibrations** are in **levels** in the same way as the **electron levels** in atoms — so you get something like a **line spectrum**.
     The **energy levels** are in **groups** though, with **very similar** energies, so you get **loads** of **lines close together** on the **spectrum**.

3)   The **fine lines** tend to **blur** together to make wide '**bands**' in the spectrum.

Spectra

This page is for Edexcel and OCR A Option 01 only.

Diffraction Gratings are Better than Prisms   *Edexcel only*

1) It is very difficult to get a big piece of **glass** (like a **prism**) pure enough. **Bubbles** and **impurities** in the glass **absorb** some of the light passing through it and can **distort** a spectrum.

2) For the very **precise** measurements of spectra that astronomers need to make, **diffraction gratings** (see p. 32) are used instead.

3) Astronomers use **reflection gratings**, which do the same kind of thing as the **transmission diffraction gratings** described on p. 32. These use **mirrors** in very thin strips, as this **reduces** any **light loss**.
(The surface of a CD or DVD acts as a reflection grating — that's why you get all the pretty colours.)

Astronomers use Absorption Spectra to work out what Stars are Made of

1) A **continuous spectrum** is produced from a **very hot** region of a star (called the **photosphere**).

2) **Absorption** by atoms in the 'atmosphere' of the star produces a characteristic **absorption spectrum**, **superimposed** on the continuous spectrum.

3) These absorption lines are called **Fraunhofer** lines. About a **million** of them have been identified in sunlight. By looking at the absorption lines from a star, the **composition** of the **stellar atmosphere** can be worked out.

The **spectrum** from the **star** is compared with **known spectra** in the lab: ⟹

Stellar spectrum (containing H, He and Na)

Hydrogen

Helium

Sodium

You can work out loads of **other stuff** from the spectra of stars, too. From the **absorption lines** and the **Doppler effect** (see p. 128) astronomers can work out **how fast** a star is **moving** away from or towards us, **how fast** it's **rotating**, whether the star has any **planets** around it etc. From the **continuous, black body** spectrum (see p. 118) you can work out a star's **temperature**, and from that its **size**, **how long** it's going to live for and what it's going to end up as (see p. 122). Quite handy then.

Practice Questions

Q1 Describe line absorption and line emission spectra. How are these two types of spectra produced?
Q2 Why do stars form line absorption spectra?
Q3 How do astronomers work out the composition of stellar atmospheres?

Exam Questions

Q1 When sunlight is passed through a diffraction grating, dark lines are seen in the spectrum.
Explain how these lines occur. [3 marks]

Q2 (a) How is the line emission spectrum for a particular element related to its line absorption spectrum? [1 mark]
(b) How is a line absorption spectrum for a particular element obtained? [3 marks]
(c) Discuss how line absorption spectra are used to determine the composition of stellar atmospheres. [2 marks]

I can sing a rainbow — but I won't, so that's alright...

Spectra are more useful to most astrophysicists than any other bit of observational data. Most of the things we know about star formation, the life cycle of stars and the nature of the Universe (phew... big stuff) has come from spectra. One of the big things at the moment is the search for extrasolar planets using spectra and the Doppler effect (the method's on p. 128).

Luminosity and Magnitude

This page is for AQA A Option 5 and OCR A Option 01 only.

There are a couple of ways to classify stars — the first is by luminosity, using the magnitude scale.

The **Luminosity** of a Star is the **Total Energy** Emitted **per Second**

1) Stars can be **classified** according to their luminosities — that is, the **total** amount of energy emitted in the form of electromagnetic radiation **each second** (see p.118).

2) The **Sun's** luminosity is about 4×10^{26} W (luminosity is measured in watts, since it's a sort of power). The **most luminous** stars have a luminosity about a **million** times that of the Sun.

3) The **intensity**, I, of an object is the power **received** from it per unit area **at Earth**. This is the effective **brightness** of an object.

Apparent Magnitude, m, is based on how **Bright** things **Appear** from **Earth**

1) The **brightness** of a star in the night sky depends on **two** things — its **luminosity** and its **distance from us** (if you ignore weather and light pollution). So the **brightest** stars will be **close** to us and have a **high luminosity**.

2) About 2000 years ago, a Greek called Hipparchus invented a way of **classifying** the **brightness** of stars (as seen from the Earth). The very **brightest** stars were given an **apparent magnitude** of **1** and the **dimmest** stars were given an apparent magnitude of **6** with the other levels catering for the stars in between.

3) In the 19th century, the scale was redefined using a strict **logarithmic** scale:

    A **magnitude 1** star has an **intensity 100 times** greater than a **magnitude 6** star.

    This means a difference of **one magnitude** corresponds to a difference in **intensity** of $100^{1/5}$ **times**. So a magnitude 1 star is about **2.5 times brighter** than a magnitude 2 star.

4) At the same time, the range was **extended** in **both directions** with the very brightest objects in the sky having **negative apparent magnitudes**.

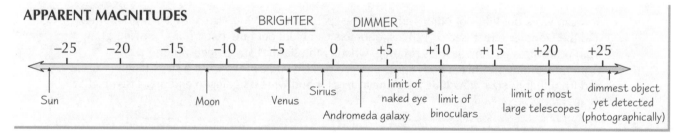

5) The **apparent magnitude**, m, is related to the **intensity**, I, by the following formula:

$$m = -2.5 \lg I + \text{constant}$$

Note that lg is $\log_{10}$, not the natural logarithm ln. That 2.5 is exactly 2.5 — not $100^{1/5}$. See next page for where it comes from.

Absolute Magnitude, M, is based only on the **Luminosity** of the Star

1) The **absolute magnitude** of a star or galaxy, M, does not depend on its distance from Earth. It is defined as what its apparent magnitude **would be** if it were **10 parsecs** away from Earth.

2) The relationship between M and m is given by the following formula:

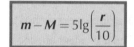

where r is the distance in parsecs (AQA A use d instead of r in this equation)

3) If you know the absolute magnitude of a star, you can use this equation to calculate its **distance** from Earth. This is really handy, since the distance to most stars is **too big** to measure using parallax (see p. 109).

    This method uses objects, such as **cepheid variable stars**, known as **standard candles**. Standard candles are objects that you can calculate the luminosity of **directly**. So, if you find a cepheid variable within a galaxy, you can work out how far that galaxy is from us. This is how the **Hubble constant** was worked out (see p. 127).

Luminosity and Magnitude

This page is for OCR A Option 01 only.

Derivation of the Absolute Magnitude Equation

If you're doing OCR A, you need to know how to derive the equation for absolute magnitude.*

1) Consider two objects, A and B, with apparent magnitudes m_A and m_B and intensities I_A and I_B.

2) From the definition of apparent magnitude, a **difference** in **magnitude** of 5 corresponds to a **factor** of 100 in **intensity**. So a difference in magnitude of $(m_A - m_B)$ corresponds to a factor of $100^{(m_A - m_B)/5}$:

$$\frac{I_B}{I_A} = 100^{(m_A - m_B)/5}$$

3) Take logs on both sides of the equation:

$$\lg\left(\frac{I_B}{I_A}\right) = \lg\left[100^{(m_A - m_B)/5}\right]$$

4) Using $\lg x^a = a \lg x$:

$$\lg\left(\frac{I_B}{I_A}\right) = \frac{(m_A - m_B)}{5}\lg 100 = \frac{2}{5}(m_A - m_B)$$

5) Rearranging:

$$m_A - m_B = 2.5\lg\left(\frac{I_B}{I_A}\right)$$

> This is just another way of writing the formula on the previous page.

6) Now imagine that objects A and B have the **same luminosity**, that A is *r* pc away and B is **10 pc** away. Since intensity reduces with distance according to the **inverse square law** (see p. 23):

$$\frac{I_B}{I_A} = \left(\frac{r}{10}\right)^2$$

7) According to the definition of absolute magnitude, if $m_A = m$ is the apparent magnitude of object A, then $m_B = M$ is its absolute magnitude, so:

$$m - M = 2.5\lg\left(\frac{r}{10}\right)^2 = 5\lg\left(\frac{r}{10}\right)$$

* This derivation uses the laws of logs which are covered fully in the AS Maths — Core 2 range.

Practice Questions

Q1 What is the relationship between apparent magnitude and intensity?

Q2 What is the equation that links apparent magnitude, absolute magnitude and distance?

Q3 Derive the equation that links *m*, *M* and *r* (no peeking now).

Exam Questions

Q1 Define the *absolute magnitude* of a star. [2 marks]

Q2 Calculate the absolute magnitude of the Sun given that the Sun's apparent magnitude is –27.
[1 pc = 2×10^5 AU] [4 marks]

Q3 The star Sirius has an apparent magnitude of –1.46 and an absolute magnitude of +1.4.
The star Canopus has an apparent magnitude of –0.72 and an absolute magnitude of –8.5.

(a) Which of the two stars appears brighter from Earth? [1 mark]

(b) Calculate the distance of Canopus from Earth. [3 marks]

Logs — the cheap and easy alternative pet...*

The magnitude scale is a pretty weird system, but like with a lot of astronomy, the old ways have stuck. Remember — the smaller the number, the brighter the object. The definition of absolute magnitude is a bit random as well — I mean, why ten parsecs? Ours not to reason why, ours but to... erm... learn it. (Doesn't have quite the same ring does it.)

Stars as Black Bodies

This page is for AQA A Option 5 only.

Now they're telling us the Sun's black. Who writes this stuff?

A **Black Body** is a **Perfect Absorber** and **Emitter**

1) Objects emit **electromagnetic radiation** due to their **temperature**. At everyday temperatures this radiation lies mostly in the **infrared** part of the spectrum (which we can't see) — but heat something up enough and it will start to **glow**.

2) **Pure black** surfaces emit radiation **strongly** and in a **well-defined way**. We call it **black body radiation**.

3) A black body is defined as:

> A body that **absorbs all wavelengths** of electromagnetic radiation (that's why it's called a **black** body) and can **emit all wavelengths** of electromagnetic radiation.

4) To a reasonably good approximation **stars** behave as **black bodies** and their black body radiation produces their **continuous spectrum** (see p. 114).

5) The graph of **intensity** against **wavelength** for black body radiation varies with **temperature**, as shown in the graph:

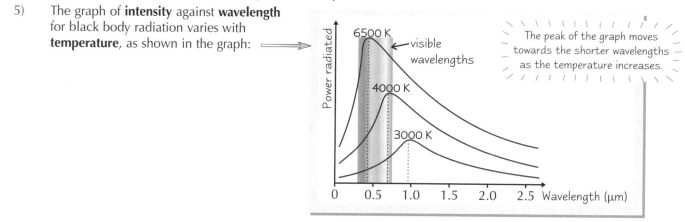

The peak of the graph moves towards the shorter wavelengths as the temperature increases.

The **Peak Wavelength** gives the **Temperature**

1) For each temperature, there is a **peak** in the black body curve at a wavelength called the **peak wavelength**, λ_{max}.

2) λ_{max} is related to the **temperature** by **Wien's displacement law**:

$$\lambda_{max}T = 0.0029 \text{ m·K}$$

where T is the temperature in kelvin and m·K is a <u>metrekelvin</u>.

The **Luminosity** of a Star Depends on its **Temperature** and **Surface Area**

1) The **luminosity** of a star is the **total energy** it emits **per second** and is related to the **temperature** of the star and its **surface area**.

2) The luminosity is proportional to the **fourth power** of the star's **temperature** and is **directly proportional** to the **surface area**. This is **Stefan's Law**:

$$L = \sigma A T^4$$

where L is the luminosity of the star (in W), A is its surface area (in m^2), T is its surface temperature in K and σ (a little Greek "sigma") is Stefan's constant.

3) Measurements give Stefan's constant as $\sigma = 5.67 \times 10^{-8} \text{ Wm}^{-2}\text{K}^{-4}$.

4) From **Earth**, we can measure the **intensity** of the star. The intensity is the **power** of radiation **per square metre**, so as the radiation spreads out and becomes **diluted**, the intensity **decreases**. If the energy has been emitted from a **point** or a **sphere** (like a star, for example) then it obeys the **inverse square law**:

$$I = \frac{L}{4\pi d^2}$$

where L is the luminosity of the star (in W), and d is the distance from the star.

Stars as Black Bodies

This page is for AQA A Option 5 and OCR A Option 01 only.

The **Atmosphere** can **Mess up** Measurements

1) **Wien's displacement law**, **Stefan's law** and the **inverse square law** can all be used by astronomers to work out various **properties** of stars. This requires very **careful measurements**, but our atmosphere mucks up the results.

2) Our atmosphere only lets **certain wavelengths** of **electromagnetic radiation** through and is **opaque** to all the others. The graph shows how the **transparency** of the atmosphere varies with **wavelength**.

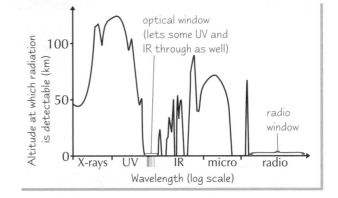

3) All this is before we even consider things like **man-made light pollution**. Observatories are placed at **high altitudes**, well away from **cities**, and in climates of **low humidity** to minimise the problem. The best solution, though, is to send up **satellites** which can take measurements **above** the atmosphere.

Our **Detectors** give Wonky Results too

1) The **measuring devices** that astronomers use aren't perfect since their **sensitivity** depends on the **wavelength**.

2) For instance, **glass absorbs UV** light but is **transparent** to **visible light**, so any instruments that use glass affect UV readings straight off.

3) **CCDs** are commonly used in careful measurements, but these are affected by the wavelength too:

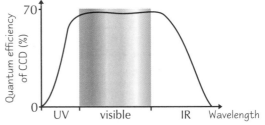

4) Astronomers get round this by **careful calibration** of their detecting instruments.

Practice Questions

Q1 What is Wien's displacement law and what is it used for?
Q2 What is the relationship between luminosity, surface area and temperature?
Q3 Why are accurate measurements of black body radiation difficult on the Earth's surface?

Exam Questions

Q1 A certain star has a surface temperature of 4000 K and has the same luminosity as the Sun (3.9×10^{26} Wm^{-2}).

(a) State the relationship between the luminosity of this star, its surface area and its temperature. [2 marks]

(b) What is the star's surface area? [2 marks]

Q2 The star Procyon A, which has a luminosity of 3.0×10^{27} Wm^{-2}, produces a black body spectrum with a peak wavelength at 530 nm.

(a) Use Wien's displacement law to estimate the surface temperature of Procyon A. [2 marks]

(b) Calculate the surface area of Procyon A. [2 marks]

Astronomy — theories, a bit of guesswork and a whole load of calibration...

Astronomy isn't the most exact of sciences, I'm afraid. The Hubble Space Telescope's improved things a lot, but try and get a look at some actual observational data. Then look at the error bars — they'll generally be about the size of your house.

Spectral Classes and the H-R Diagram

This page is for AQA A Option 5 only.

As well as classifying stars by luminosity (the magnitude scale, p.116), they can be classified by colour.

The **Visible** Part of **Hydrogen's Spectrum** is called the **Balmer Series**

1) The lines in **emission** and **absorption spectra** (see p.114) occur because electrons in an atom can only exist at certain well-defined **energy levels** (see p. 66).

2) In **atomic hydrogen**, the electron is usually in the **ground state** ($n = 1$) but there are lots of energy levels ($n = 2$ to $n = \infty$ — called excitation levels) that the electron **could** exist in if it was given more energy.

> The wavelengths corresponding to the **visible bit** of hydrogen's spectrum are caused by electrons moving from **higher energy levels** to the **first excitation level** ($n = 2$). This leads to a series of **lines** called the **Balmer series**.

The **Strengths** of the **Spectral Lines** Show the **Temperature** of a Star

1) For a **hydrogen absorption line** to occur in the **visible** part of a star's spectrum (see page 114), electrons in the hydrogen atoms already need to be in the $n = 2$ state.

2) This happens at **high temperatures**, where **collisions** between the atoms can give the electrons the extra energy they need.

3) If the temperature is **too high**, though, the majority of the electrons will reach the $n = 3$ level instead, which means there won't be so many Balmer transitions.

4) So the **intensity** of the Balmer lines depends on the **temperature** of the star.

5) For a particular intensity of the Balmer lines, **two temperatures** are possible. Astronomers get around this by looking at the **absorption lines** of other atoms and molecules as well.

The **Relative Strength** of Absorption Lines gives the **Spectral Class**

Well... quite.

1) For historical reasons the stars are classified into:

> **spectral classes: O** (hottest), **B, A, F, G, K** and **M**

Use a **mnemonic** to remember the order. The standard one is the rather non-PC '**Oh Be A Fine Girl, Kiss Me**'.

2) The graph shows how the **intensity** of the **visible spectral lines** changes with **temperature**:

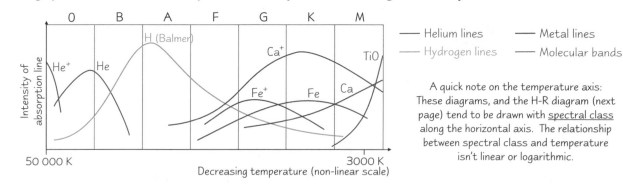

A quick note on the temperature axis: These diagrams, and the H-R diagram (next page) tend to be drawn with <u>spectral class</u> along the horizontal axis. The relationship between spectral class and temperature isn't linear or logarithmic.

> **THE SPECTRAL CHARACTERISTICS OF SPECTRAL CLASSES**
>
> i) **O** — blueish-white stars. The strongest visible spectral lines are **helium absorptions**, since these need a really high temperature.
>
> ii) **B, A** — white stars. The visible spectra are governed by the Balmer **hydrogen absorption** lines.
>
> iii) **F, G** — yellowish stars. These spectra are governed by **metal absorptions**.
>
> iv) **K, M** — reddish stars. **Molecular bands** (see p. 114) are present in the spectra of these stars, since they're cool enough for molecules to form.

3) These spectral classes are **subdivided** into **O0** to **O9** then **B0** to **B9** and so on. This means that a star's **classification** gives its **temperature** to within **5%**. The **Sun** has a temperature of **5800 K** and is a **G2** star.

Spectral Classes and the H-R Diagram

This page is for AQA A Option 5 and OCR A Option 01 only.

Absolute Magnitude vs Temperature/Spectral Class — the H-R diagram

1) Independently, Hertzsprung and Russell noticed that a plot of **absolute magnitude** (see p. 116) against **temperature** (or **spectral class**) didn't just throw up a random collection of stars but showed **distinct areas**.

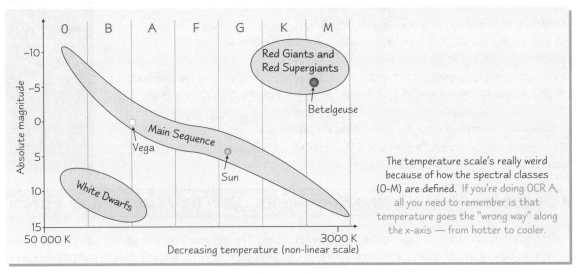

The temperature scale's really weird because of how the spectral classes (O-M) are defined. If you're doing OCR A, all you need to remember is that temperature goes the "wrong way" along the x-axis — from hotter to cooler.

2) The **long, diagonal band** is called the **main sequence**. Main sequence stars are in their long-lived **stable phase** where they are fusing **hydrogen** into **helium**. The Sun is a main sequence star.

3) Objects that have a **high luminosity** and a relatively **low surface temperature** must have a **huge** surface area because of Stefan's law (page 118). These stars are called **red giants** and are found in the **top-right** corner of the H-R diagram. Red giants are stars that have **moved off** the main sequence and fusion reactions other than hydrogen to helium are also happening in them.

4) Objects that have a **low luminosity** but a **high temperature** must be very **small**, again because of Stefan's law. These stars are called **white dwarfs** and are about the size of the Earth. They lie in the **bottom-left** corner of the H-R diagram. White dwarfs are stars at the **end** of their lives, where all of their fusion reactions have stopped and they are just **slowly cooling down**.

Practice Questions

Q1 Why does hydrogen have to be at a particular temperature before Balmer absorption lines are seen?

Q2 List the spectral classes in order of decreasing temperature and outline their spectral characteristics.

Q3 What is an H-R diagram and what are the three main groups of stars that emerge when the diagram is plotted?

Exam Questions

Q1 (a) Describe how the Balmer series is formed in a hydrogen absorption spectrum. [4 marks]

(b) Explain how temperature affects the strength of the Balmer lines in stellar absorption spectra. [3 marks]

(c) In which two spectral classes of star are strong Balmer lines observed? [2 marks]

Q2 The spectra of K and M stars have absorption bands corresponding to energy levels of molecules. Explain why this only occurs in the lowest temperature stars. [2 marks]

Q3 Draw the basic features of an H-R diagram indicating where one would find main sequence stars, red giants and white dwarfs. [5 marks]

'Ospital Bound — A Furious Girl's Kicked Me...

Spectral classes are another example of astronomers sticking with tradition. The classes used to be ordered alphabetically by the strength of the Balmer lines. When astronomers realised this didn't quite work, they just fiddled around with the old classes rather than coming up with a sensible new system. Just to make life difficult for people like you and me.

Stellar Evolution

*This page is for AQA A Option 5 **and** OCR A Option 01 only.*

Stars go through several different stages in their lives and move around the H-R diagram as they go (see p. 121).

Stars *Begin as Clouds of* Dust *and* Gas

1) Stars are born in a **cloud** of **dust** and **gas**, most of which was left when previous stars blew themselves apart in **supernovae**. The denser clumps of the cloud **contract** (very slowly) under the force of **gravity**.

2) When these clumps get dense enough, the cloud fragments into regions called **protostars**, that continue to contract and **heat up**.

3) Eventually the **temperature** at the centre of the protostar reaches a **few million degrees**, and **hydrogen nuclei** start to **fuse** together to form helium (see page 94).

4) This releases an **enormous** amount of **energy** and creates enough **pressure** (radiation pressure) to stop the **gravitational collapse**.

5) The star has now reached the MAIN SEQUENCE and will stay there, relatively **unchanged**, while it fuses hydrogen into helium.

Cloud of Dust and Gas

Protostar

Main Sequence Star

Main Sequence *Stars become* Red Giants *when they* Run Out *of* Fuel

1) Stars spend most of their lives as **main sequence** stars. The **pressure** produced from **hydrogen fusion** in their **core balances** the **gravitational force** trying to compress them. This stage is called **core hydrogen burning**.

2) When the **hydrogen** in the **core** runs out nuclear fusion **stops**, and with it the **outward pressure stops**. The core **contracts** and **heats up** under the **weight** of the star.

3) The material **surrounding** the core still has **plenty of hydrogen**. The **heat** from the contracting **core** raises the **temperature** of this material enough for the hydrogen to **fuse**. This is called **shell hydrogen burning**. *(Very low-mass stars stop at this point. They use up their fuel and slowly fade away...)*

4) The core continues to contract until, eventually, it gets **hot** enough and **dense** enough for **helium** to **fuse** into **carbon** and **oxygen**. This is called **core helium burning**. This releases a **huge** amount of energy, which **pushes** the **outer layers** of the star outwards. These outer layers **cool**, and the star becomes a RED GIANT.

5) When the **helium** runs out, the carbon-oxygen core **contracts again** and heats a **shell** around it so that helium can fuse in this region — **shell helium burning**.

Low Mass *Stars (like the Sun)* Eject *their Shells leaving behind a* White Dwarf

1) The **carbon-oxygen core isn't hot enough** for any further **fusion** and so it continues to **contract** under its own **weight**. Once the core has shrunk to about **Earth-size**, **electrons** exert enough pressure to stop it collapsing any more. This is due to a quantum effect called **electron degeneracy** (fret not, you don't have to know about this).

2) The **helium shell** becomes more and more **unstable** as the core contracts. The star **pulsates** and **ejects** its outer layers into space as a **planetary nebula**, leaving behind the dense core.

3) The star is now a very **hot**, **dense solid** called a WHITE DWARF, which will simply **cool down** and **disappear**.

High Mass *Stars have a* Shorter Life *and a more* Exciting Death

1) Even though stars with a **large mass** have a **lot of fuel**, they use it up **more quickly** and don't spend so long as main sequence stars.

2) When they are **red giants** the 'core burning to shell burning' process can continue beyond the fusion of helium, building up layers in an **onion-like structure**. For **really massive** stars this can go all the way up to **iron**.

3) Nuclear fusion **beyond iron** isn't **energetically favourable**, though, so once an iron core is formed then very quickly it's goodbye star.

4) The star explodes cataclysmically in a SUPERNOVA, leaving behind a NEUTRON STAR or (if the star was massive enough) a BLACK HOLE.

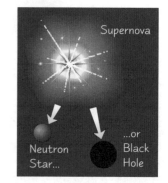

Supernova

Neutron Star...

...or Black Hole

Stellar Evolution

This page is for AQA A Option 5 *only.*

Massive Stars *go out with a* Bit of a Bang

1) When the core of a star runs out of fuel, it starts to **contract** — forming a white dwarf core.

2) If the star is **massive enough**, though, **electron degeneracy** can't keep the core intact. This happens when the mass of the core is more than **1.4 times** the mass of the **Sun**.

3) The electrons get **squashed** onto the atomic **nuclei**, combining with protons to form **neutrons** and **neutrinos**.

4) The core collapses to become a **NEUTRON STAR**, but the outer layers are still **falling** towards it rapidly.

5) When the outer layers **hit** the surface of the **neutron star** they **rebound**, setting up huge **shockwaves**, ripping the star apart and causing a **supernova**. The light from a supernova can outshine an **entire galaxy**!

> Neutron stars are incredibly **dense** (about 4×10^{17} kgm^{-3}).
>
> They're **very small**, typically about 20 km across, and they can **rotate very fast** (up to 600 times a second).
>
> They emit **radio waves** in two beams as they rotate. These beams sometimes sweep past the Earth and can be observed as **radio pulses** rather like the flashes of a lighthouse. These rotating neutron stars are called **PULSARS**.

Neutron Stars *are* Weird *but* Black Holes *are a* Lot Weirder

1) If the **core** of the star is more than **3 times** the **Sun's mass**, the **neutrons** can't withstand the gravitational forces.

2) There are **no known mechanisms** left to stop the core collapsing to an **infinitely dense** point called a **singularity**. At that point, the **laws of physics** break down completely.

3) Up to a certain distance away (called the **Schwarzschild radius**) the gravitational pull is **so strong** that nothing, not even **light**, can escape its grasp. The **boundary** of this region is called the **event horizon**.

> <u>The Schwarzschild radius is the distance at which the escape velocity is the speed of light</u>
>
> An object moving at the **escape velocity** has **just enough kinetic energy** to overcome the black hole's gravitational field.
>
> From Newton's law of gravitation we get $\frac{1}{2}mv^2 = \frac{GMm}{r}$ where m = mass of object, M = mass of black hole, v = velocity of object, r = distance from black hole
>
> Dividing through by m and making r the subject gives: $r = \frac{2GM}{v^2}$
>
> By replacing v with the speed of light, c, you get the Schwarzschild radius, R_s: $\boxed{R_s \simeq \frac{2GM}{c^2}}$.

This formula is just an <u>approximation</u> though — Newton's law of gravity doesn't quite work in intense gravitational fields, so Einstein's <u>general relativity</u> should be used instead (the maths is way too hard though — see p. 138).

Practice Questions

Q1 Outline how the Sun was formed, how it will evolve and how it will die.

Q2 Describe a white dwarf and a neutron star. What are the main differences between them?

Exam Questions

Q1 Outline the main differences between the evolutions of high mass and low mass stars, starting from when they first become main sequence stars. [6 marks]

Q2 (a) What is meant by the Schwarzschild radius of a black hole? [2 marks]

(b) Calculate an estimate of the Schwarzschild radius for a black hole that has a mass of 6×10^{30} kg. [2 marks]

(c) Why does the formula only give an approximate value for the Schwarzschild radius? [1 mark]

Live fast — die young...

The more massive a star, the more spectacular its life cycle. The most massive stars burn up the hydrogen in their core so quickly that they only live for a fraction of the Sun's lifetime — but when they go, they do it in style.

Galaxies and Quasars

This page is for OCR A Option 01 only. (But if you're doing AQA A Option 5 you might want to read the bit on our galaxy.)

Not all galaxies are the same — you get elliptical ones and spiral ones and ring-shaped ones, and there's that one shaped like an enormous... Anyway, the only galaxy you really need to know about in detail is our own Milky Way.

Our Galaxy is the **Milky Way**

1) The Milky Way is a **barred spiral galaxy** (see below) about **100,000 ly** in diameter.

2) Surrounding the Milky Way there's a **halo**, which consists of **very old** stars (called population II) and **globular clusters** (large spherical groups of stars of about the same age).

3) The **disk** contains **spiral arms** of **young stars** (called population I) as well as interstellar **dust** and **gas**.

4) The **central bulge** (with both young and old stars at quite a high density) contains the **galactic nucleus**. This is a source of strong **radio emissions** and many astronomers believe this to be a **supermassive black hole**.

5) The **Sun** lies about **28,000 ly out** from the centre of the galaxy in the **Orion arm**. This is a **short**, **stubby** spiral arm between two major arms.

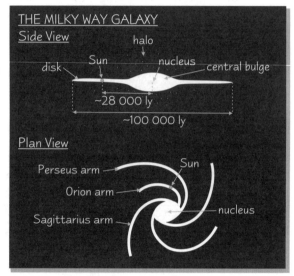

Galaxies come in all **Shapes** and **Sizes**

1) At the beginning of the 20th century, Edwin Hubble showed that some of the '**fuzzy patches**' in the night sky were actually **separate galaxies** of stars a long distance from the Milky Way.

2) In 1925, he **classified** their **shapes** into **spiral**, **barred spiral**, **elliptical** and **irregular**. He also **subdivided** these classes depending on things like the **relative size** of the spiral arms, or the **roundness** of the elliptical galaxies.

3) A typical **giant galaxy** (like the Milky Way) has a diameter of about **100,000 ly**. Dwarf elliptical galaxies (which are quite common) are about **500 ly** in diameter whereas **giant elliptical galaxies** (much, much rarer) can be up to **2,000,000 ly** in diameter.

The **Mass** of a Galaxy can be Worked Out from **How Fast** its Stars Orbit it

1) A galaxy **doesn't** spin with a **constant angular speed** — the **individual stars** within it orbit at **varying speeds**.

2) The **speed of orbit** depends on the **mass** of the galaxy **within the orbit** and the **distance** the star is from the **galactic centre**. The formula for working out the mass of the galaxy within a star's orbit can be derived from **Newton's law of gravitation**:

The force of **gravity** on the star provides the **centripetal force** (see p. 13), so:

$$\frac{GMm}{r^2} = \frac{mv^2}{r}$$

where m is the mass of the star, v is the speed of the star, r is the distance of the star from the centre of the galaxy and M is the mass of the galaxy within the star's orbit.

Simplifying and **rearranging** to make M the subject gives the formula: $\boxed{M = \dfrac{v^2 r}{G}}$

Example   The Sun orbits the galactic centre at a speed of 230 kms⁻¹ and a distance of 2.6 × 10²⁰ m. Calculate the mass of the Milky Way within the Sun's orbit.

$$M = \frac{\left(2.3 \times 10^5\right)^2 \times 2.6 \times 10^{20}}{6.67 \times 10^{-11}} = 2.1 \times 10^{41} \text{ kg}$$

That's about 100 billion times the mass of the Sun.

Galaxies and Quasars

This page is for AQA A Option 5 only.

Quasars — Quasi-Stellar Objects

1) **Quasars** were discovered in 1960 and were first thought to be **stars in our galaxy**.

2) The puzzling thing was that their spectra were **nothing like** normal stars. They sometimes shot out **jets** of material, and many of them were very active **radio sources**.

3) The 'stars' produced a **continuous spectrum** that was nothing like a blackbody radiation curve and instead of absorption lines, there were **emission lines** of elements that astronomers **had not seen before**.

4) However, these lines looked strangely familiar and in 1963 Maarten Schmidt realised that they were simply the **Balmer series** of hydrogen but **redshifted** enormously.

Quasars are a Very Long Way Away so they must be Very Bright

1) Pages 127 and 128 talk about this in detail, but the **huge redshift** means that these quasars are **moving away** from us at **great speed**. That suggests they're a **huge distance away** — in fact, the **most distant** objects yet seen.

2) The measured redshifts give us distances of **billions of light years**. Using the **inverse square law** for intensity (see p. 23) gives an idea of just how **bright** quasars are:

> *Example*   A quasar has about the same intensity as a star 4000 ly away with the same luminosity as the Sun (4×10^{26} W). Its redshift gives a distance of 2×10^{10} ly. Calculate its luminosity.
>
> $I_{quasar} = I_{star}$ so they cancel out of the equation. $L \propto Id^2 \Rightarrow \dfrac{L_{quasar}}{L_{star}} = \dfrac{d_{quasar}^{\,2}}{d_{star}^{\,2}} \Rightarrow L_{quasar} = L_{star} \cdot \dfrac{d_{quasar}^{\,2}}{d_{star}^{\,2}} = 4 \times 10^{26} \cdot \dfrac{4 \times 10^{20}}{16 \times 10^{6}} = 1 \times 10^{40}$ W
>
> That's bright — over **3000 times** the **luminosity** of the **entire Milky Way galaxy**!

3) And there's very good evidence to suggest that quasars are only about the size of our **solar system**.

4) Let me run that past you again. **That's the power of 10^{13} Suns from something the size of the Solar System.**

5) These numbers caused a lot of controversy in the astrophysics community — they seemed crazy. Many astrophysicists thought there must be a more reasonable explanation. But then evidence for the distance of quasars came when **sensitive CCD** equipment detected the fuzzy cloud of **a galaxy around a quasar**.

> The current consensus is that a quasar is a very powerful **galactic nucleus** — a **supermassive**, spinning **black hole** (about 100 million times the mass of the Sun) at the centre of a distant galaxy. This black hole is surrounded by a doughnut shaped mass of **whirling gas** falling into it, which produces the light. In the same way as a pulsar (see p.123), magnetic fields produce jets of radiation streaming out from the poles. The black hole must consume about **10 Suns per year** to produce the energy observed.

Practice Questions

Q1   Outline the Hubble system for classifying galaxies.

Q2   What evidence is there that quasars are not stars in our galaxy?

Q3   Sketch a side view and a plan view of our galaxy, outlining the main features. Indicate the approximate position of the Sun.

Exam Questions

Q1   A star was found to be orbiting the circumference of a circular galaxy at a speed of 500 kms^{-1}. If the radius of the galaxy is 50 000 ly, calculate its mass.   [3 marks]

Q2   (a) What evidence is there to suggest that quasars are a very long distance away?   [1 mark]

     (b) Use the concept of the inverse square law to suggest why quasars must be very bright.   [2 marks]

     (c) What do astronomers believe to be the main features of a quasar?   [2 marks]

Long ago, in a galaxy far, far away — there was a radio-loud, supermassive black hole with a highly luminous acr...

Quasars are weird. There's still some disagreement in the astrophysics community about what they are. There's even some evidence (not generally accepted) that quasars are much nearer than the redshift suggests and are just moving very quickly.

Models of the Universe

This page is for OCR A Option 01 only.

Right, we're moving on to the BIG picture now — we all like a bit of cosmology...

Copernicus put the Sun at the Centre of the Universe

1) Until the **17th century**, the accepted model of the Universe was the one described by **Ptolemy**. He placed **Earth** at its **centre**, surrounded by **crystal spheres** containing the **Sun, Moon, planets** and **stars**.

2) It fitted many of the facts but didn't match the **motion** of the **planets**. The main problem is that some of the planets seem to go **backwards** during part of their orbit — called **retrograde motion**.

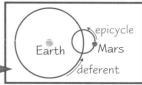

3) Ptolemy solved this by splitting up the orbits of the planets into **deferents** and **epicycles** (this is a bit like the orbit of the Moon around the Sun).

4) Ptolemy's model still **didn't quite work** and later astronomers had to keep tweaking it.

5) In **1504**, **Copernicus** developed a **much simpler** model that put the **Sun** at the centre of the Universe. This is called a **heliocentric model**.

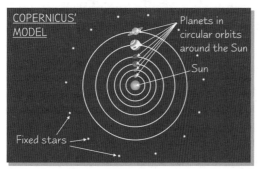

Copernicus' model **wasn't widely accepted**. The **catholic church** didn't like it on **theological grounds** and there were **scientific problems** too.

The **main problem** was that the model still **didn't quite match** the **motion** of the **planets**. It was a much closer approximation than in the previous models, but those clumsy **epicycles** had to be reintroduced for it to match observations exactly.

Kepler Refined Copernicus' Model and Galileo found Evidence to Support it

1) In **1609**, **Kepler** published his **three laws of planetary motion** (see page 36). He realised that Copernicus' model worked **perfectly** if the planets move in **elliptical orbits** around the Sun instead of **circular ones**.

2) At about the same time **Galileo** was studying the night sky with a **telescope**. He discovered that:

- the **'heavens'** weren't perfect — there were **mountains** on the **Moon** and **spots** on the **Sun**
- the **Milky Way** was made up of millions of **very faint stars**
- Venus shows **phases** like the **Moon, consistent** with the **heliocentric model**
- **Jupiter** has **moons** of its own. There were **heavenly bodies** orbiting other bodies than the **Earth**.

Galileo got into **a lot of trouble** with the **Church** about this, but eventually his ideas became **widely accepted**.

An Infinite Universe leads to Olber's Paradox

1) The **demotion** of **Earth** from anything special is taken to its logical conclusion with the **Cosmological Principle**...

COSMOLOGICAL PRINCIPLE: on a **large scale** the Universe is **homogeneous** (every part is the same as every other part) and **isotropic** (everything looks the same in every direction) — so it doesn't have a **centre**.

2) Until the **1930s**, cosmologists believed that the Universe was **infinite** in both **space** and **time** (that is, it had always existed) and **static**. This was the **only way** that it could be **stable** using **Newton's law** of gravitation. Even **Einstein modified** his theory of **general relativity** to make it consistent with the **Steady-State Universe**.

3) In 1826, though, an astronomer called **Olber** had noticed a **big problem** with this model of the Universe.

If stars (or galaxies) are **spread randomly** throughout an **infinite** Universe then **every possible line of sight** must contain a **star**. Calculations show that this should make the **whole** night sky **uniformly bright**.

This problem is called **Olber's paradox** and clearly there's a **contradiction** with an infinite and static Universe.

Models of the Universe

This page is for Edexcel, AQA A Option 5, OCR A Option 01 and OCR B only.

Hubble *Realised that the* Universe *was* Expanding

1) The **spectra** (see p. 114) from **galaxies** (apart from a few very close ones) all show **redshift**. That is, the **characteristic spectral lines** of the elements are all at a longer wavelength than you would expect them to be. The amount of **redshift** gives the **recessional velocity** — how fast the galaxy is moving away (see page 128).

2) A plot of **recessional velocity** against **distance** (found using cepheid variables — see p. 116) showed that they were **proportional**, which suggests that the Universe is **expanding**. This gives rise to **Hubble's law**:

$$v = H_0 d$$

where v = recessional velocity in **kms^{-1}**, d = distance in **Mpc** and H_0 = **Hubble's constant** in **kms^{-1}Mpc^{-1}**.

3) Since distance is very difficult to measure, astronomers disagree on the value of H_0. It's generally accepted that H_0 lies somewhere between 50 kms^{-1}Mpc^{-1} and 100 kms^{-1}Mpc^{-1}.

4) The **SI unit** for H_0 is s^{-1}. To get H_0 in SI units, you need v in ms^{-1} and d in m (1 Mpc = 3.09×10^{22} m).

The Expanding Universe *gives rise to the* Hot Big Bang Model

1) The Universe is **expanding** and **cooling down** (because it's a closed system — see p. 74). So further back in time it must have been **smaller** and **hotter**. If you trace time back **far enough**, you get a **Hot Big Bang**...

> **THE HOT BIG BANG THEORY**: the Universe started off as an **infinitely hot**, **infinitely dense** point (a singularity) and has been **expanding** ever since.

2) Since the Universe is **expanding uniformly** away from **us** it seems as though we're at the **centre** of the Universe, but this is an **illusion**. You would observe the **same thing** at **any point** in the Universe.

The Age *and* Observable Size *of the* Universe *Depend on* H_0

1) If the Universe has been **expanding** at the **same rate** for its whole life, the **age** of the Universe is $t = 1/H_0$ (time = distance/speed). This is only an estimate since the Universe probably hasn't always been expanding at the same rate.

2) Unfortunately, since no one knows the **exact value** of H_0 we can only guess the Universe's age. If H_0 = 75 kms^{-1}Mpc^{-1}, then the age of the Universe $\approx 1/(2.4 \times 10^{-18}$ s$^{-1}) = 4.1 \times 10^{17}$ s = **13 billion years**.

3) The **absolute size** of the Universe is **unknown** but there is a limit on the size of the **observable Universe**. This is simply a **sphere** (with the Earth at its centre) with a **radius** equal to the **maximum distance** that **light** can travel during its **age**. So if H_0 = 75 kms^{-1}Mpc^{-1} then this sphere will have a radius of **13 billion light years**.

4) This gives a very simple solution to **Olber's paradox**. If the observable Universe is **finite**, then there is **absolutely no reason** why every line of sight should include a star. Actually, most of them don't.

Practice Questions

Q1 Outline Copernicus' model of the Universe.

Q2 State Olber's paradox. How does the Hot Big Bang model resolve it?

Exam Questions

Q1 Explain the roles of Kepler and Galileo in the acceptance of the heliocentric model of the Universe. [4 marks]

Q2 (a) State Hubble's law, explaining the meanings of all the symbols. [2 marks]

(b) What does Hubble's law suggest about the nature of the Universe? [2 marks]

(c) Assume H$_0$ = 50 kms^{-1}Mpc^{-1} (1 Mpc = 3.09×10^{22} m).

i) Calculate H$_0$ in SI units. [2 marks]

ii) Calculate an estimate of the age of the Universe, and hence the size of the observable Universe. [3 marks]

Try telling THAT to the Inquisition...

So there are no deferents and epicycles, hmmmm? Iiiiiiiiiinteresting...   *Bring out... the comfy chair!*

Evidence for the Hot Big Bang

This page is for Edexcel, AQA A Option 5, OCR A Option 01 and OCR B only.

The Hot Big Bang theory (HBB to his mates) is pretty much accepted now (with a few tweaks) — make sure you know why.

The **Red Shift** of Galaxies is **Strong Evidence** for the HBB

1) Hubble realised that the **speed** that **galaxies moved away** from us depended on **how far** they were away. This led to the idea that the Universe started as an **infinitely hot**, **infinitely dense** point and is currently **expanding** (the Hot Big Bang; see page 127).

2) The **speed** of a particular galaxy can be calculated by measuring the **redshift** of the light coming from it. **Redshift** is similar to a phenomenon called the **Doppler effect**.

The **Doppler Effect** — the **Motion** of a Wave's **Source** Affects its **Wavelength**

1) You'll have experienced the Doppler effect **loads of times** with **sound waves**.

2) Imagine an ambulance driving past you. As it moves **towards you** its siren sounds **higher-pitched**, but as it **moves away**, its **pitch** gets **lower**. This change in **frequency** and **wavelength** is called the **Doppler shift**.

3) The frequency and the wavelength **change** because the waves **bunch together** in **front** of the source and **stretch out behind** it. The **amount** of stretching or bunching together depends on the **velocity** of the **source**.

4) When a **light source** moves **away** from us, the wavelengths of its light become **longer** and the frequencies become lower. This shifts the light towards the **red** end of the spectrum and is called **redshift**.

5) When a light source moves **towards** us, the **opposite** happens and the light undergoes **blueshift**.

6) The amount of redshift or blueshift is determined by the following formula:

$$z = \frac{\Delta\lambda}{\lambda} = \frac{v}{c} \quad \text{if } v \ll c$$

z is the redshift, $\Delta\lambda$ is the difference between the observed wavelength and the emitted wavelength, λ is the emitted wavelength, v is the velocity of the source in the direction of the observer and c is the speed of light. ($v \ll c$ means "v is much less than c".)

7) The way cosmologists tend to look at this stuff, the galaxies aren't actually moving **through space** away from us. Instead, **space itself** is expanding and the light waves are being **stretched** along with it. This is called **cosmological redshift** to distinguish it from **redshift** produced by sources that **are** moving through space.

8) The same formula works for both types of redshift as long as v is much less than c. If v is close to the speed of light, you need to use a nasty, relativistic formula instead (you don't need to know that one).

Redshift is Used to Study Spectroscopic **Binary Stars**   *AQA A Option 5 only*

1) About **60%** of the stars we observe are actually **two stars** that orbit each other. Most of them are too far away from us to be **resolved** with **telescopes** but the **lines** in their **spectra** (see p.114) show a binary star system. These are called **spectroscopic binary stars**.

2) By observing how the **absorption lines** in the spectrum change with **time** the **period** of orbit can be calculated:

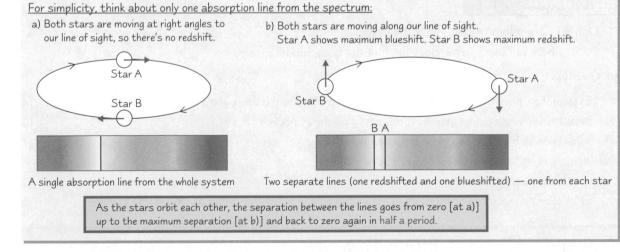

3) Astronomers use a similar method to find **extrasolar planets**.

Evidence for the Hot Big Bang

This page is for OCR A Option 01 and OCR B only.

Cosmic Microwave Background Radiation — More Evidence for the HBB

1) The Hot Big Bang model predicts that loads of **electromagnetic radiation** was produced in the **very early Universe**. This radiation should **still** be observed today (it hasn't had anywhere else to go).
2) Because the Universe has **expanded**, the wavelengths of this cosmic background radiation have been **stretched** and are now in the **microwave** region.
3) This was picked up **accidentally** by Penzias and Wilson in the 1960s.

Properties of the Cosmic Microwave Background Radiation (CMBR)

1) In the late 1980s a satellite called the **Cosmic Background Explorer** (**COBE**) was sent up to have a **detailed look** at the radiation.
2) It found a **perfect blackbody spectrum** corresponding to a **temperature of 2.73 K** (see page 118).
3) The radiation is **isotropic** and **homogeneous**, which confirms the Cosmological Principle (see page 126).
4) There are **very tiny fluctuations** in temperature, which are at the limit of detection. These are due to tiny energy-density variations in the early Universe, and are needed for the initial 'seeding' of galaxy formation.
5) The background radiation also shows a **Doppler shift**, indicating the Earth's motion through space. It turns out that the **Milky Way** is rushing towards an unknown mass (the **Great Attractor**) at over a **million miles an hour**.

Another Bit of Evidence is the Amount of Helium in the Universe

1) The HBB model also explained the **large abundance of helium** in the Universe (which had puzzled astrophysicists for a while).
2) The early Universe had been very hot, so at some point it must have been hot enough for **hydrogen fusion** to happen. This meant that, together with the theory of the synthesis of the **heavier elements** in stars, the **relative abundances** of all of the elements could be accounted for.

Practice Questions

Q1 What is the Doppler effect?
Q2 Write down the formula for the redshift and blueshift of light.
Q3 Explain how the spectra of binary stars can be used to calculate their period of orbit.
Q4 What is the cosmic background radiation?

Exam Questions

Q1 (a) A certain object has a redshift of 0.37. Estimate the speed at which it is moving away from us. [2 marks]
   (b) Use Hubble's law to estimate the distance (in light years) that the object is from us.
       (Take $H_0 = 2.4 \times 10^{-18}$ s^{-1}, 1 ly = 9.5×10^{15} m.) [2 marks]
   (c) With reference to the speed of the object, explain why your answers to a) and b) are estimates. [1 mark]

Q2 The spectra of three objects have been taken. What can you deduce from each of the following?
   (a) The absorption lines from object A have been shifted towards the blue end of the spectrum. [1 mark]
   (b) The absorption lines from object B have been shifted towards the red end of the spectrum. [1 mark]
   (c) The absorption lines from object C oscillate either side of their normal position in the spectrum with a period of two weeks. [2 marks]

Q3 Describe the main features of the cosmic background radiation and explain why its discovery was considered strong evidence for the Hot Big Bang model of the Universe. [6 marks]

Neeeaaaaaaaaaawwwwwwww...

The simple Big Bang model doesn't actually work — not quite, anyway. There are loads of little things that don't quite add up. Modern cosmologists are trying to improve the model using a period of very rapid expansion called inflation.

Evolution of the Universe

These pages are for OCR A Option 01 *only.*

Are you sitting comfortably... then I'll begin.

The **Properties** of the **Early Universe** are Investigated in **Particle Accelerators**

1) The early Universe was **very**, **very hot** and so anything that was present would have had **a lot of energy**.

2) Experimenters can **recreate** these kinds of energies in **particle accelerators** by **smashing** beams of particles (like protons and electrons) into either **fixed targets** or **antimatter beams** moving in the opposite direction (see p.102).

3) At the point of **collision**, temperatures of over a **billion K** can be created.

4) This corresponds to the temperature of the Universe within its first **millisecond**. There is **no direct experimental evidence** for the conditions of the Universe before this time.

This is what the **Experiments** have Told us up to Now

1) **Matter** comes in **two types**: **quarks** and **leptons** (see pages 96 to 101 for more details).

2) At the sorts of energies that exist in the current Universe, quarks are **never** on their **own** but group together to form particles like **protons** and **neutrons**.

3) **Leptons** are a family of particles including **electrons** and **neutrinos**.

4) There are **four fundamental forces**, each affecting different types of particle.

The standard model of particle physics is covered in a lot more detail in Section 9.

Type of Interaction	Particles Affected
strong	quarks only (e.g. protons and neutrons)
electromagnetic	charged particles only
weak	all types
gravity	all types

It's hard to predict what the Universe was like before 0.01 seconds

Of the four forces, the strong force was dominant in the Universe up to 0.01s after the Big Bang. The strong force isn't very well understood, and it's incredibly hard to predict. So as well as having very little in the way of experimental evidence, the theories are a bit thin on the ground at the moment as well.

Matter and *Antimatter* are Produced in *Equal* Amounts (ish)

1) According to Einstein, **energy** and **mass** are **equivalent** — $E = mc^2$. That means, given enough **energy**, you can create **particles**. Whenever we see this today, the energy is converted into a **particle-antiparticle pair**. For example, a gamma photon can turn into an **electron** and a **positron** (see p. 99).

2) If **matter** and **antimatter** were made in equal amounts (**matter-antimatter symmetry**) in the **early Universe**, they would have annihilated each other completely, leaving **just electromagnetic radiation**. We're here, so clearly that didn't happen.

For every **billion antimatter particles** produced in the early Universe, **a billion and one matter particles** were produced. When the particles annihilated, they left behind a small excess of matter.

GUTs try to Explain the Very *Beginning* of the *Universe*

1) A **Grand Unified Theory** (**GUT**) attempts to **unify** the four different forces into a **single force**.

2) The idea is that at **very high temperatures** there isn't any **distinction** between the forces and that at **low temperatures** they're all just different **aspects** of the same force.

3) The **electromagnetic** and nuclear **weak** forces have already been **unified** into the **electroweak force**. (Although a predicted particle, the Higgs boson, hasn't been found yet.)

4) GUTs have had some success in unifying the **strong force** as well, but none have been generally accepted yet.

5) The biggest stumbling block is gravity. We need a theory of **quantum gravity** to explain the earliest Universe.

Evolution of the Universe

The Story So Far...

Before 10^{-4} seconds after the Big Bang, this is mainly guesswork, really. There are plenty of theories out there, but not much experimental evidence to back them up. The general consensus at the moment goes something like this:

1) **Big Bang to 10^{-43} seconds.** Well, it's anybody's guess, really. At this sort of size and energy, even General Relativity stops working properly. This is the "infinitely hot, infinitely small, infinitely dense" bit.

2) **10^{-43} seconds to 10^{-4} seconds.** The Universe expands and cools. The unified force splits up into gravity, nuclear strong, nuclear weak and finally electromagnetic forces. Many cosmologists believe the Universe went through a rapid period of expansion called inflation at about 10^{-34} s. The Universe is a sea of quarks, antiquarks, leptons and photons. The quarks aren't bound up in particles like protons and neutrons, because there's too much energy around.

   At some point, matter-antimatter symmetry gets broken (see previous page). Nobody knows exactly how or when this happened, but most cosmologists like to put it as early as possible in the history of the Universe (before inflation, even).

Now we're onto more solid ground

3) **10^{-4} seconds.** This corresponds to a temperature of 10^{12} K. The Universe is cool enough for quarks to join up to form particles like protons and neutrons. They can never exist separately again. Matter and antimatter annihilate each other, leaving a small excess of matter and huge numbers of photons (resulting in the cosmic background radiation that we observe today).

4) **About 100 seconds.** Temperature has cooled to 10^9 K. The Universe is similar to the interior of a star. Protons are cool enough to fuse to form helium nuclei.

5) **About 300 000 years.** Temperature has cooled to about 3000 K. The Universe is cool enough for electrons (that were produced in the first millisecond) to combine with helium and hydrogen nuclei to form atoms. The Universe becomes transparent since there are no free charges for the photons to interact with. This process is called recombination.

6) **About 15 billion years (now).** Temperature has cooled to about 3 K. Slight density fluctuations in the Universe mean that, over time, clumps of matter have been condensed by gravity into galactic clusters, galaxies and individual stars.

Practice Questions

Q1 Why is there no experimental evidence for the physics involved before the very early Universe?
Q2 Why is it difficult to develop theories describing the evolution of the universe before 0.01 seconds?
Q3 What is recombination?

Exam Questions

Q1 (a) What is meant by matter-antimatter symmetry? [2 marks]
   (b) Describe the inconsistency of this phenomenon with the observed Universe. [2 marks]
   (c) How do some theories resolve this problem? [1 mark]
   (d) What do cosmologists believe is the source of cosmic background radiation? [1 mark]

Q2 Starting from the production of matter and antimatter, describe the evolution of the Universe (including its structure) up to the present day. [10 marks]

In the beginning...

Don't worry too much about the particle physics bits — you don't need to understand the standard model in detail. Just remember the matter-antimatter symmetry problem and the bit on GUTs and you'll be fine.

Fate of the Universe

This page is for AQA B (just this bit), Edexcel and OCR A Option 01 only.

This page assumes the Standard Big Bang Model — that is, the Universe started from an infinitely dense point, expanded smoothly to the size it is now, and is still expanding. So we can ignore newfangled theories like inflation.

Gravity Warps Space and Time

1) **General relativity** explains that gravity works by changing the **shape** of space and time (see p. 138)

2) To reduce the brain-ache a bit, you can imagine the Universe as a **2-dimensional surface** that's warped in 3-dimensions. This is a handy way of getting an idea of what's going on, but **be careful**. **Space-time** actually has **4-dimensions** (x, y, z, and time).

3) On a big scale, there are **three ways** that gravity can warp the Universe: the Universe can be **flat**, **open**, or **closed**.

4) This **curvature** of space-time determines the eventual **fate** of the Universe.

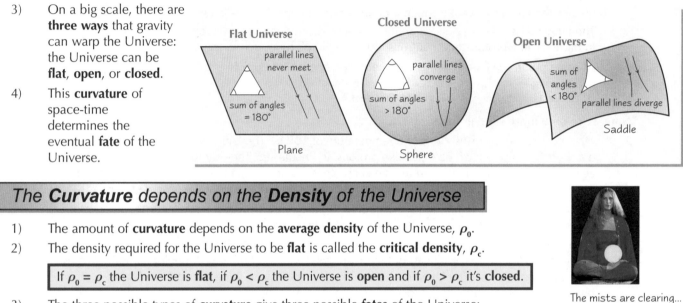

Flat Universe — parallel lines never meet — sum of angles = 180° — Plane

Closed Universe — parallel lines converge — sum of angles > 180° — Sphere

Open Universe — sum of angles < 180° — parallel lines diverge — Saddle

The Curvature depends on the Density of the Universe

1) The amount of **curvature** depends on the **average density** of the Universe, ρ_0.

2) The density required for the Universe to be **flat** is called the **critical density**, ρ_c.

> If $\rho_0 = \rho_c$ the Universe is **flat**, if $\rho_0 < \rho_c$ the Universe is **open** and if $\rho_0 > \rho_c$ it's **closed**.

The mists are clearing...

3) The three possible types of **curvature** give three possible **fates** of the Universe:

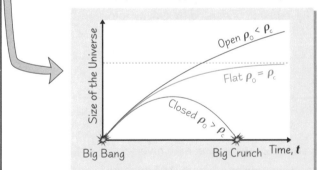

Size of the Universe / Time, t — Open $\rho_0 < \rho_c$ — Flat $\rho_0 = \rho_c$ — Closed $\rho_0 > \rho_c$ — Big Bang — Big Crunch

In an **open** Universe, gravity (controlled by the density) is **too weak** to stop the expansion. The Universe will just keep **expanding forever**.

In a **closed** Universe, gravity is **strong enough** to stop the expansion and start the Universe **contracting** again (ending up with a **Big Crunch**).

In a **flat** Universe, gravity is **just strong enough** to stop the expansion at $t = \infty$ (so the Universe expands forever, but more and more slowly with time).

We Can't Calculate the Age of the Universe until we know its Density

1) A reasonable **estimate** of the **age** of the Universe is found from $t \approx 1/H_0$ (see page 127). But this formula assumes that the Universe had been expanding at the **same rate** for its whole lifetime.

2) In fact if you look at the **graph** of size against time, the **expansion rate** of the Universe is **slowing down**, even for the **open Universe**. So **in the past** the Universe was expanding **faster** than it is now.

3) That means we've **overestimated** the time it's taken for the Universe to get to the size it is now.

4) The **more dense** the Universe is, the **younger** it must be.

5) If you include all the "**dark matter**" that's been detected **indirectly**, current **estimates** of the actual density of the Universe aren't far off the **critical density**.

In fact, if you project the Universe **back in time**, the current estimate would agree with the critical density to **thirty decimal places**!

6) Since Physics is based on **elegance** and **simplicity** (stop laughing), a lot of cosmologists believe that $\rho_0 = \rho_c$ — but that's just a hunch.

Fate of the Universe

This page is for OCR A Option 01 only.

You can Work Out the Density of the Universe from its Escape Velocity

1) **Assume** the Universe is **flat** and that it obeys **Newtonian mechanics**.

2) Based on these assumptions, you can derive the **density** of the **Universe** using **Newton's law of gravitation**. (Of course, the Universe **doesn't** obey Newtonian mechanics — a **strict derivation** needs **general relativity**.)

3) You can **model** the Universe as an **expanding sphere**, with radius **R**, mass **M** and a constant density ρ_0.

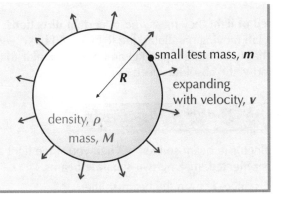

density, ρ_0
mass, **M**

small test mass, **m**

expanding with velocity, **v**

4) To model a **flat** Universe, the **surface** of the sphere must be moving at the **escape velocity** of the gravitational field.

5) Imagine a small test mass, **m**, on the surface of the sphere. If it has **just enough** kinetic energy to escape the gravitational field, then:

$$\frac{1}{2}mv^2 = \frac{GMm}{R} \quad \text{... equation 1}$$

6) The sphere is expanding at a rate given by **Hubble's law**, so $v = H_0 R$ (see p. 127).

7) **Substitute** for **v** in equation 1, and after a bit of **simplifying** and **rearranging** you get: $R^3 = \dfrac{2GM}{H_0^2}$ ... equation 2

8) The density, ρ_0, of the sphere is **mass/volume**, and the volume of the sphere is $\dfrac{4}{3}\pi R^3$.

Substitute for **R** in equation 2 and tinker around a bit and you get the density formula: $\boxed{\rho_0 = \dfrac{3H_0^2}{8\pi G}}$

Example Calculate the density of the Universe if H_0 = 75 kms^{-1}Mpc^{-1}.

$H_0 = 2.4 \times 10^{-18}$ s^{-1}

So $\rho_0 = \dfrac{3 \times [2.4 \times 10^{-18}]^2}{8\pi \times 6.67 \times 10^{-11}} = \textbf{1} \times \textbf{10}^{-26}$ **kgm^{-3}**

See p. 127 for how to convert the Hubble constant into SI units

---

Practice Questions

Q1 What are the three possible fates of the Universe?

Q2 Why does the calculated age of the Universe depend on its density?

Q3 Derive the formula for calculating the density of the Universe.

Exam Questions

Q1 (a) Cosmologists believe the Universe to be flat. What evidence is there to suggest that this is the case? [3 marks]

(b) Explain what is meant by a flat universe in terms of its geometry, and its evolution. [3 marks]

Q2 The upper limit on Hubble's constant is 100 kms^{-1}Mpc^{-1} (1 Mpc = 3.09×10^{22} m).

(a) Work out the average density of the Universe if the Universe is flat and H_0 is 100 kms^{-1}Mpc^{-1}. [4 marks]

(b) Given that the mass of a hydrogen atom is 1.7×10^{-27} kg, calculate the average number of hydrogen atoms in every m^3 of space if the entire mass of the Universe is hydrogen. [2 marks]

It's the end of the world as we know it...

Recently, astronomers have found evidence that the expansion rate is actually <u>accelerating</u>. That means the simple shapes of the graphs on the last page might be a long way from the true picture... but you just pretend you didn't read that. ☺

The Speed of Light & Relativity

These pages are for AQA A Option 8 and OCR A Option 01 only.

First — a bit of a history lesson.

Michelson and Morley tried to find the Absolute Speed of the Earth

1) During the 19th century, most physicists believed in the idea of **absolute motion**. They thought everything, including light, moved relative to a **fixed background** — something called the **aether**.

2) **Michelson** and **Morley** tried to measure the **absolute speed** of the **Earth** through the aether using a piece of apparatus called an **interferometer**.

3) They expected the motion of the Earth to affect the **speed of light** they measured in **certain directions**. According to Newton, the speed of light measured in a **lab** moving parallel to the light would be $(c + v)$ or $(c - v)$, where v is the speed of the lab. By measuring the speed of light **parallel** and **perpendicular** to the motion of the Earth, Michelson and Morley hoped to find v, the absolute speed of the Earth.

They used an Interferometer to Measure the Speed of the Earth

The interferometer was basically **two mirrors** and a **partial reflector** (a beam-splitter). When you shine light at a partial reflector, some of the light is **transmitted** and the rest is **reflected**, making **two separate beams**.

The mirrors were at **right angles** to each other, and an **equal distance**, L, from the beam-splitter.

The Michelson-Morley Interferometer

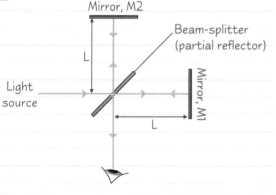

1) **Monochromatic light** is sent towards the **beam-splitter**.
2) The light is split into **two beams** travelling at **right angles** to each other.
3) The beams are reflected at **mirrors M1** and **M2**.
4) When the reflected beams meet back at the beam-splitter, they form an **interference pattern** (see p. 30 – 31).
5) This interference pattern is **recorded** by the observer.
6) Then the whole interferometer is **rotated** through **90°** and the experiment **repeated**.

> **EXPECTED OUTCOME**
>
> According to Newton's laws, light moving parallel to the motion of the Earth should take longer to travel to the mirror and back than light travelling at right angles to the Earth's motion. So rotating the apparatus should have changed the travel time for the two beams.
>
> This would cause a tiny shift in the interference pattern.

They Didn't get the Result they were Expecting

They **repeated** the experiment **over** and **over** again — at different **times of day** and at different points in the **year**. Taking into account any **experimental errors**, there was **absolutely no shift** in the interference pattern.

The time taken by each beam to travel to each mirror was **unaffected** by rotating the apparatus.

So, Newton's laws **didn't work** in this situation.

Most scientists were really puzzled by this "null result". Eventually, the following **conclusions** were drawn:

a) It's **impossible** to detect **absolute motion** — the aether doesn't exist.

b) The **speed of light** has the **same value** for all observers.

The Speed of Light & Relativity

The **invariance** of the speed of the light is one of the cornerstones of special relativity. The other is based on the concept of an **inertial frame of reference**.

Anything Moving with a *Constant Velocity* is in an *Inertial Frame*

A reference frame is just a **space** that we decide to use to locate the **position of an object** — you can think of a reference frame as a **set of coordinates**.

> An **inertial reference frame** is one in which **Newton's 1st law** is obeyed. (Newton's 1st law says that objects won't accelerate unless they're acted on by an external force.)

1) Imagine sitting in a carriage of a train **waiting at a station**. You put a **marble** on the table. The marble **doesn't move**, since there aren't any horizontal **forces** acting on it. **Newton's 1st law** applies, so it's an **inertial frame**.

2) You'll get the **same result** if the carriage moves at a **steady speed** along a **smooth straight track** — another inertial frame.

3) As the train **accelerates** out of the station, the marble **moves** without any force being applied. Newton's 1st law **doesn't apply**. The accelerating carriage **isn't an inertial frame**.

4) **Rotating** or **accelerating** reference frames **aren't** inertial. In most cases, though, you can think of the **Earth** as an inertial frame — it's near enough.

A stationary train carriage is an inertial frame

An accelerating train carriage is NOT an inertial frame

Einstein's *Postulates* of *Special Relativity*

Einstein's theory of **special relativity** only works in **inertial frames** and is based on **two postulates** (assumptions):

1) **Physical laws have the same form in all inertial frames.**
2) **The speed of light in free space is invariant.**

1) The first postulate says that if we do **any physics experiment** in any inertial frame we'll always get the **same result**. That means it's **impossible** to use the result of **any experiment** to work out if you're in a **stationary reference frame** or one moving at a **constant velocity**.

2) The second postulate says that the **speed of light** (in a vacuum) always has the **same value**. It isn't affected by the **movement** of the **person measuring it** or by the movement of the **light source**.

Practice Questions

Q1 Draw a labelled diagram showing the apparatus used to determine the absolute speed of the Earth. Include the light source, mirrors, beam-splitter and the position of the observer.

Q2 State the postulates of Einstein's theory of special relativity.

Q3 Explain why a carriage on the London Eye is not a inertial frame.

Exam Questions

Q1 In the Michelson-Morley interferometer experiment, interference fringes were observed. When the apparatus was rotated through 90 degrees the expected result was not observed.

(a) What was the expected result? [1 mark]

(b) What conclusions were eventually drawn from the null observation? [2 marks]

Q2 (a) Using a suitable example, explain what is meant by an inertial reference frame. [2 marks]

(b) Explain what is meant by the invariance of the speed of light. [2 marks]

The speed of light is always the same — whatever your reference frame...

Michelson and Morley showed that Newton's laws didn't always work. This was a <u>huge</u> deal. Newton's laws of motion had been treated like gospel by the physics community since the 17th century. Then along came Herr Einstein...

Special Relativity

These pages are for AQA A Option 8 and OCR A Option 01 only.

Special relativity ONLY WORKS IN INERTIAL FRAMES — it doesn't work in an accelerating frame.

A Moving Clock Runs Slow

1) Time runs at **different speeds** for two observers **moving relative** to each other.

2) A **stationary** observer measures the interval between two events as t_0, the **proper time**. (Since there's no such thing as absolute motion, a "stationary" observer means someone that's stationary relative to the reference frame the events are happening in.) An observer moving at a **constant velocity**, v, will measure a **longer** interval, t, between the two events. t is given by the equation:

$$t = \frac{t_0}{\sqrt{1 - \frac{v^2}{c^2}}}$$

where $\sqrt{1 - \frac{v^2}{c^2}}$ is called the relativity factor and c is the speed of light.

3) This is called **time dilation**.

> **A THOUGHT EXPERIMENT TO ILLUSTRATE TIME DILATION**
>
> *Anne is on a high-speed train travelling at 0.9c. She switches on a torch for exactly 2 seconds.*
> *Claire is standing on the platform and sees the same event, but records a longer time.*
> *It appears to Claire that Anne's clock is running slow.*
>
> In this experiment, **Anne** is the **stationary observer**, so she measures the **proper time**, t_0.
> Claire is **moving at 0.9c relative to the events**, and so measures a time t given by:
>
> $$t = \frac{t_0}{\sqrt{1 - \frac{v^2}{c^2}}} = \frac{2}{\sqrt{1 - \frac{(0.9c)^2}{c^2}}} = \frac{2}{\sqrt{1 - 0.9^2}} = 4.59 \text{ s}$$
>
> To the **external observer** (e.g. Claire) **moving clocks** run **slowly**.

It's really important that you get the "stationary observer" right.

There's Proof of Time Dilation from Muon Decay

1) **Muons** are **particles** created in the **upper atmosphere** that move towards the ground at speeds close to c.

2) In the laboratory (**at rest**) they have a **half-life** of less than **2 μs**. From this half-life, you would expect most muons to **decay** between the top of the atmosphere and the Earth's surface, but that **doesn't happen**.

Experiment to Measure Muon Decay

1) Measure the speed, v, of the muons (this is about **0.99c**).
2) Place a detector (MR1) at high altitude and measure the muon count rate.
3) Use another detector (MR2) to measure the count rate at ground level.
4) Compare the two figures.

Here are some typical results:
MR1 = 500 per minute
MR2 = 290 per minute
Distance between detectors (d) = 2000 m
Time as measured by an observer = d/v = 6.73 μs
Half-life of muons at rest = 1.53 μs

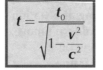
muons moving with velocity, v
MR1
d
MR2

3) We can do some calculations using the data above. In the reference frame of the **observer** the muons seemed to have travelled for **4.4 half-lives** between the two detectors. You would expect the count rate at the **second detector** to be only about **25 counts per minute**.

4) However, in a **muon's reference frame**, travelling at 0.99c, the time taken for the journey is just $t_0 = 0.94$ μs. From the point of view of the muons, the time elapsed is **less** than their **half-life**. From the point of view of the observer, it appears that the half-life of the muons has been **extended**.

Special Relativity

A Moving Rod Looks Shorter

1) A **rod** moving in the **same direction** as its **length** looks **shorter** to an external observer.

2) A **stationary** observer measures the length of an object as l_0. An observer moving at a **constant velocity**, v, will measure a shorter length, l. l is given by the equation:

$$l = l_0\sqrt{1 - \frac{v^2}{c^2}}$$

This is called **length contraction**.

A THOUGHT EXPERIMENT TO ILLUSTRATE LENGTH CONTRACTION

Anne (still in the train moving at 0.9c) measures the length of her carriage as 3 m. Claire, on the platform, measures the length of the carriage as it moves past her.

Claire measures a length: $l = l_0\sqrt{1 - \frac{v^2}{c^2}} = 3\sqrt{1 - \frac{(0.9c)^2}{c^2}} = 3\sqrt{1 - 0.9^2} = 1.3\,\text{m}$

The Mass of an Object Increases with Speed

1) The **faster** an object **moves**, the **more massive** it gets.

2) An object with rest mass m_0 moving at a **velocity** v has a **relativistic mass** m given by the equation:

$$m = \frac{m_0}{\sqrt{1 - \frac{v^2}{c^2}}}$$

So, near the speed of light, increasing an object's <u>kinetic energy</u> increases its <u>mass</u>.

3) As the relative speed of an object approaches c, the mass approaches **infinity**. So, in practice, no massive object can move at a speed **greater than** or **equal to** the speed of light.

Mass and Energy are Equivalent

1) Einstein extended his idea of **relativistic mass** to write down the most famous equation in physics: $\boxed{E = mc^2}$

2) This equation says that **mass** can be **converted** into **energy** and vice versa. Or, alternatively, **any energy** you supply to an object **increases** its **mass** — it's just that the increase is usually **too small** to measure.

3) The **total energy** of a relativistic object is given by the equation: ⟹ $E = \frac{m_0 c^2}{\sqrt{1 - \frac{v^2}{c^2}}}$

This is just substituting the relativistic mass into $E = mc^2$.

Practice Questions

Q1 State the equations for time dilation and length contraction, carefully defining each symbol.

Q2 Using the results from the muon experiment (page 136), show that the time elapsed in the reference frame of the muon is 0.94 μs.

Q3 A particle accelerated to near the speed of light gains a very large quantity of energy. Describe how the following quantities change as the particle gains more and more energy: a) the mass; b) the speed.

Exam Questions

Q1 A subatomic particle has a half-life of 20 ns when at rest. If a beam of these particles is moving at $0.995c$ relative to an observer, calculate the half-life of these particles in the frame of reference of the observer. [3 marks]

Q2 Describe a thought experiment to illustrate time dilation. [4 marks]

Q3 For a proton ($m_0 = 1.6 \times 10^{-27}$ kg) travelling at 2.8×10^8 ms^{-1} calculate:
 (a) the relativistic mass, [1 mark]
 (b) the total energy. [1 mark]

Have you ever noticed how time dilates when you're revising physics...

*In a moving frame, time stretches out, lengths get shorter and masses get bigger. One of the trickiest bits is remembering which observer's which — t_0, m_0 and l_0 are the values you'd measure if the object was **at rest**.*

General Relativity

These pages are for *OCR A Option 01* **only**

You think this stuff is hard — just be glad you don't have to see the maths... *shudder*

General Relativity *is Based on the* Principle *of* Equivalence

1) The **general theory of relativity** (**GR**) is a theory of **gravity** based on the **principle of equivalence**. It extends Einstein's treatment of motion to reference frames that **aren't inertial** — accelerating frames.

> **PRINCIPLE OF EQUIVALENCE:** From inside a closed laboratory, it's **impossible** to tell the **difference** between the effect of a **gravitational field** and the effect of the **laboratory accelerating**.

2) So you can't do **any physics experiment** (from inside the room) that could tell you whether you were in a room on the **surface of the Earth**, or in a spacecraft **accelerating at 9.81 ms⁻²**.

3) General relativity predicts that gravitational fields will affect not only masses, but **time** and **light**.

Clocks *Run* Slow *Near* Massive Objects

1) General relativity predicts that the **time interval** between two events as measured in a **strong gravitational field** will be **longer** than the interval between the same events as measured in a **weaker gravitational field**.

> **A THOUGHT EXPERIMENT TO ILLUSTRATE THE EFFECT OF GRAVITY ON TIME**
>
> An observer in an **inertial** frame of reference observes **clock A** that's **accelerating** away from him. The acceleration stretches out the **interval** between each "**tick**" reaching the observer from **clock A** — compared to the clock in his hand. **A** runs slower.
>
> **Clock B** is accelerating away from him **faster** than clock A, so the effect will be more pronounced.
>
>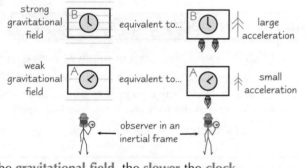
>
> The **faster** the clock **accelerates** away from the observer, the **slower** it **runs** (relative to the observer).
>
> Applying the **principle of equivalence**, the **stronger** the **gravitational field**, the **slower** the **clock** — so a clock on the ground floor of a very tall building on Earth will run slower than one on the top floor.

2) This effect has been **detected** by launching an **atomic clock** to **10 000 km** and recording the change to the flow of time (a difference of just 0.16 μs per hour). The results agreed very well with predictions from GR.

Gravity *can* Bend Light

According to the general theory of relativity, **light** will **curve** in a gravitational field.

> **A THOUGHT EXPERIMENT TO ILLUSTRATE THE BENDING OF LIGHT IN A GRAVITATIONAL FIELD**
>
> Imagine an **accelerating glass spacecraft** (i.e. one you can see through).
>
> A **beam of light** passes through the glass spacecraft at **right angles** to its motion. In the **time** it takes for the light to cross the space, the **velocity** of the spacecraft has **increased**. So an observer **in** the craft will see the light **bend** downwards in a **parabola**.
>
> An observer **at rest** outside the craft will see the light travel in a **straight line** across the spacecraft.
>
>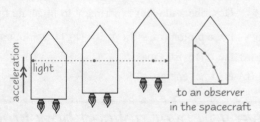
>
> Applying the **principle of equivalence**, light should curve in a **gravitational field**.

General Relativity

Evidence from the 1919 Solar Eclipse Supported GR

1) In **1919**, two teams of scientists travelled to **West Africa** and **Brazil** to photograph a **total solar eclipse** (that's when the Moon blocks out most of the Sun's light).

2) Stars that should have been **completely hidden** behind the **Sun** appeared in the photograph. The light from these stars must have been **curved** by the Sun's gravitational field.

3) The **amount** of deflection **agreed very closely** with the predictions from the general theory of relativity.

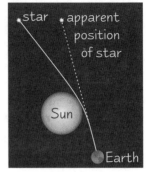

Mercury's Orbit also Supports General Relativity

1) All the planets orbit the Sun in **elliptical** orbits. The point in the orbit where the planet is **closest** to the **Sun** is called the **perihelion**.

2) **Mercury's orbit** is slightly peculiar. It's been known since 1845 that the **position** of Mercury's **perihelion changes** very slightly on each orbit. This effect is called **precession**.

3) The perihelion takes nearly a **quarter of a million years** to get back to its starting point.

4) The **rate of precession** couldn't be explained using Newton's laws, but calculations based on general relativity agree very well with the observations. (This result came from solving Einstein's tensor field equations. Anyone up for a bit of tensor calculus? No?)

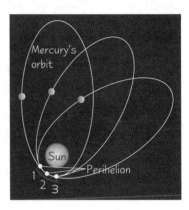

To detect GR effects, you need **very rapid accelerations**, **strong gravitational fields** or **extremely accurate clocks**. But not all its effects are quite so "grand scale" as the ones on this page.

General relativity is used in **Global Positioning Satellite** (GPS) systems. They can only pinpoint locations as accurately as they do by taking into account the effects of general relativity.

Practice Questions

Q1 What type of reference frame is equivalent to an inertial frame in a gravitational field?

Q2 Whose laws failed to accurately explain the orbit of Mercury?

Q3 What is meant by the word "perihelion"?

Q4 Which clock would run faster, a clock on the top of Mt. Everest or one at sea level?

Q5 Using diagrams, describe a thought experiment to demonstrate the effect of a gravitational field on the path of a light ray.

Exam Questions

Q1 Using diagrams, explain the principle of equivalence. [3 marks]

Q2 Explain how Mercury's orbit lends support to the general theory of relativity. [3 marks]

Q3 During the 1919 solar eclipse, observers photographed stars that should not have been visible. Describe the results, and explain their significance for the general theory of relativity. [3 marks]

You can pick your friends...

General relativity is one of the most successful theories of all time (along with quantum mechanics). It explains everything from gravitational lensing (where a big mass magnifies a distant object by bending the light like a lens) to black holes. It doesn't agree with quantum mechanics on small scales though, so there's still some work to be done. Any takers?

Body Mechanics

These pages are for OCR A Option 02 only.

And here you were, thinking you were doing physics — looks like biology to me...

The **Body's** held up by the **Skeleton**

1) The **skeleton protects** sensitive **organs** and forms a **rigid frame** that holds the body together.

2) **Joints** in the skeleton give **freedom of movement**.

3) **Ligaments** are bands or sheets of **fibrous tissue** that keep the bones in the **right place** while the joint's moving.

4) **Synovial fluid** inside some joints acts as a **lubricant** to reduce friction.

5) The **ends** of some bones are covered in **cartilage**, which acts as a **shock absorber**.

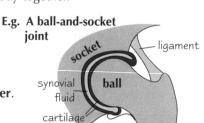

E.g. A ball-and-socket joint

Muscles are **Connected** to the **Bones** by **Tendons**

1) **Skeletal muscles** are bundles of specialised body tissue that **contract** when they receive a **nerve signal**.

2) **Tendons** are **strong** cords that are **flexible** but **inelastic** (don't stretch) and **connect** the muscles to the bones.

3) A muscle can **only contract**, and so can **pull** in **one direction** only. So, for **control** of a joint, muscles have to be used in **pairs** that **oppose** each other — a **flexor** and an **extensor** — which make an **antagonistic pair**.

Muscles, **Bones** and **Joints** Act as **Levers**

1) In a lever, an **effort force** (from the muscle) acts against a **load force** (e.g. the weight of your arm) by means of a **rigid object** (the bone) rotating around a pivot or **fulcrum** (the joint).

2) The **mechanical advantage** (**M.A.**) of a lever is the **load force** divided by the **effort force**.

$$M.A. = \frac{load}{effort}$$

3) Most of the **joints** of the **human body** are **class 3 levers** — they have an M.A. of a **lot less** than 1.

4) You can use the **principle of moments** to solve lever questions:

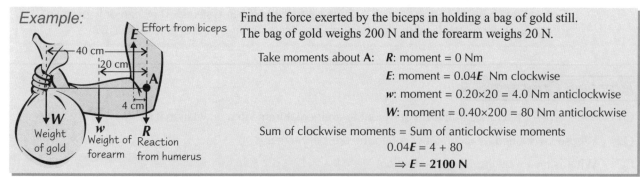

Example:

Effort from biceps

W — Weight of gold
w — Weight of forearm
R — Reaction from humerus

Find the force exerted by the biceps in holding a bag of gold still. The bag of gold weighs 200 N and the forearm weighs 20 N.

Take moments about **A**:   **R**: moment = 0 Nm

E: moment = 0.04**E** Nm clockwise

w: moment = 0.20×20 = 4.0 Nm anticlockwise

W: moment = 0.40×200 = 80 Nm anticlockwise

Sum of clockwise moments = Sum of anticlockwise moments

0.04**E** = 4 + 80

⇒ **E** = **2100 N**

Bending and **Lifting Increase** the **Forces** on your **Spine**

1) When you're standing, your upper body exerts forces on the **lumbosacral disc**, a fibrous **pad** between the **sacrum** and the **bottom lumbar vertebra**.

2) The spine is **curved**, so the weight of your upper body (about **60%** of your total weight) acts on the disc at an angle, causing a **compressive** force and a **shear** force.

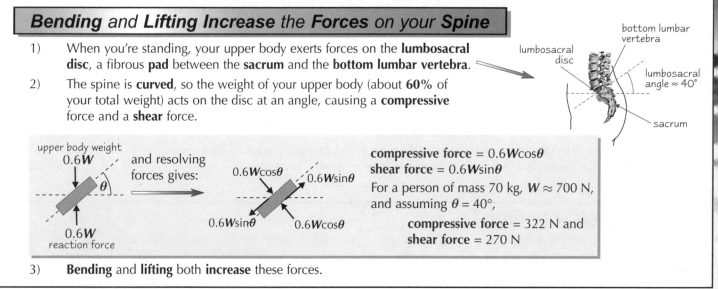

and resolving forces gives:

compressive force = 0.6**W**cosθ
shear force = 0.6**W**sinθ

For a person of mass 70 kg, **W** ≈ 700 N, and assuming θ = 40°,

    compressive force = 322 N and
    shear force = 270 N

3) **Bending** and **lifting** both **increase** these forces.

Body Mechanics

Good Posture Reduces the Chances of Pain and Damage

1) The lumbosacral disc is particularly **susceptible** to damage by **shear forces**. **Bad posture** often results in the **pelvis** tilting **forward**, **increasing** the lumbosacral angle, θ, and hence the **shear force**. A person bending their upper body through 45° exerts a force on the disc roughly five times as much as when standing upright. ⟹

2) **Bending** over and/or **lifting** with **straight legs** massively **increases** the **shear force** and can lead to a 'slipped disc'.

3) To **reduce** the risk of damage when **lifting**, keep your **back vertical** and **bend your knees**.

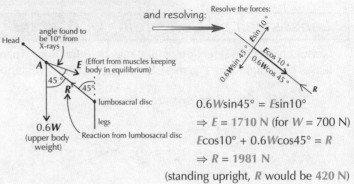

and resolving:

Resolve the forces:

$0.6W\sin45° = E\sin10°$
$\Rightarrow E = 1710$ N (for $W = 700$ N)
$E\cos10° + 0.6W\cos45° = R$
$\Rightarrow R = 1981$ N
(standing upright, R would be **420 N**)

Standing, Walking and Running

1) When you **stand still**, the **ground** exerts a **normal reaction force**, R (which is **equal** and **opposite** to your **weight**).

2) When you **walk** or **run**, your leg **accelerates** upwards and forwards as it lifts off the ground and **decelerates** as it lands back on the ground.

3) The **force** that produces these **accelerations** and **decelerations** is called the **ground force**, G. It's a combination of the **normal reaction force**, R (greater than your weight this time) and the **frictional force**, f.

4) The **maximum possible** value of f is given by $f = \mu R$, where μ is the **coefficient of friction** between your shoes and the ground. This limits the **length** your **stride** can be without slipping.

5) G acts **along the leg** and its magnitude **increases** with **speed**.

Standing
$G = R$

Decelerating
$G^2 = R^2 + f^2$

Accelerating
$G^2 = R^2 + f^2$

Practice Questions

Q1 Describe the mechanical functions of bones, tendons, ligaments, joints and muscles.

Q2 What is an antagonistic pair of muscles, and why is a pair of muscles needed?

Q3 Draw a diagram indicating the forces involved when someone lifts a heavy object without bending their knees. Why should they bend their knees and keep their back as vertical as possible?

Exam Questions

Q1 When a person is standing, 35% of their body weight acts vertically through each hip joint. Calculate the compressive force and the shear force that act on the femoral head for a person of weight 700 N.

0.35W
50°
Femoral head

[3 marks]

Q2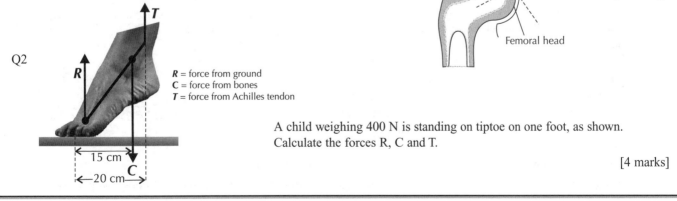

R = force from ground
C = force from bones
T = force from Achilles tendon

A child weighing 400 N is standing on tiptoe on one foot, as shown. Calculate the forces R, C and T.

[4 marks]

15 cm
20 cm

SIT UP STRAIGHT — yes, you...

Makes your back ache just thinking about it. This stuff might look scary, but it's just the simple mechanics you did at GCSE and AS. Don't let all the biology-speak put you off — when you get down to it, it's all just levers and reaction forces.

The Physics of Vision

These pages are for AQA A Option 6 and OCR A Option 02 only.

*The eyes contain **converging lenses** which focus light rays to form images.*

Lenses can be Converging or Diverging

Lenses change the direction of light rays by **refraction**. There are two types of lens. **Converging** (or **convex**) lenses bring light rays together. **Diverging** (or **concave**) lenses spread light rays out.

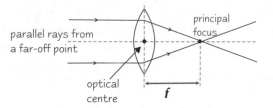

CONVEX LENS: light rays are converged

parallel rays from a far-off point

principal focus

optical centre

f

CONCAVE LENS: light rays are diverged

optical centre

principal axis

principal focus

f

1) The distance between the **optical centre** of the lens and the **principal focus** is called the **focal length**, *f*.

2) The **power**, *P*, of a lens depends on its **focal length**: The units of *P* are **dioptres** (D) (if *f* is in metres).

3) A more **powerful** lens has a shorter focal length and bends light rays more strongly.

$$P = \frac{1}{f}$$

If you put several lenses together, the total power of the lens system is just the sum of the individual powers.

Real is Positive, Virtual is Negative

(see p. 110 for more on real and virtual images)

Lenses can produce **real** or **virtual** images, and you need to follow the "**real** is **positive**, **virtual** is **negative**" rule.

1) A **converging lens** produces a **real image**, so its **focal length**, *f*, is **positive**.

2) A **diverging lens** produces a **virtual image**, so it has a **negative focal length**.

3) The focal length, *f*, is related to object distance, *u*, and image distance, *v*, by the **lens equation**...

$$\frac{1}{f} = \frac{1}{u} + \frac{1}{v}$$

4) The **linear magnification** of a lens is $m = \dfrac{\text{size of image}}{\text{size of object}}$ and $m = \dfrac{v}{u}$

You Need to Know the Basic Structure of the Eye

1) The **cornea** is a **transparent** 'window'. It has a **convex** shape, and a **high refractive index** at its interface with the air. (You did refractive index at AS.) The cornea does most of the eye's focusing.

2) The **aqueous humour** is a **watery** substance that lets light pass through the pupil to the lens.

3) The **iris** is the coloured part of the eye. It consists of **radial** and **circular muscles** that control the size of the **pupil** — the hole in the middle of the iris. This regulates the intensity of light entering the eye.

4) The **lens** acts as a **fine focus** and is controlled by the **ciliary muscles**. When the ciliary muscles **contract**, tension is released and the lens takes on a **fat**, more **spherical** shape. When they **relax**, the **suspensory ligaments** pull the lens into a **thin, flatter** shape.

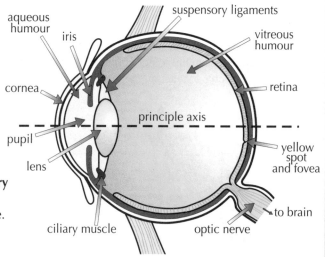

aqueous humour

iris

cornea

pupil

lens

ciliary muscle

suspensory ligaments

vitreous humour

retina

principle axis

yellow spot and fovea

to brain

optic nerve

5) The **vitreous humour** is a **jelly-like** substance that keeps the eye's shape.

6) Images are formed on the **retina**, which contains **light sensitive cells** called **rods** and **cones** (see p.144).

7) The **yellow spot** is a particularly sensitive region of the retina, on the eye's **principal axis**. In the centre of the yellow spot is the **fovea**. This is the part of the retina with the highest concentration of **cones**.

8) The **optic nerve** carries signals from the rods and cones to the **brain**.

The Physics of Vision

The **Eye** is an **Optical Refracting System**

1) The **cornea** and **aqueous humour** act as a **fixed converging lens** with a **power** of about **41 D**.

2) The power of the eye's **lens** itself is about **18 D** when viewing far-off points. By changing shape, it can increase to about **29 D** in young people when viewing near points, but can hardly increase at all in old people (see p.146).

3) You can **add together** the **powers** of the cornea, aqueous humour and lens. That means you can think of the eye as a **single converging lens** of **power 59 D** (when viewing far-off points). This gives a **focal length** of **1.7 cm**.

4) When looking at near objects, the eye's power **increases**, as the lens changes shape and the **focal length decreases** — but the distance between the lens and the image, **v**, stays the same, at 1.7 cm.

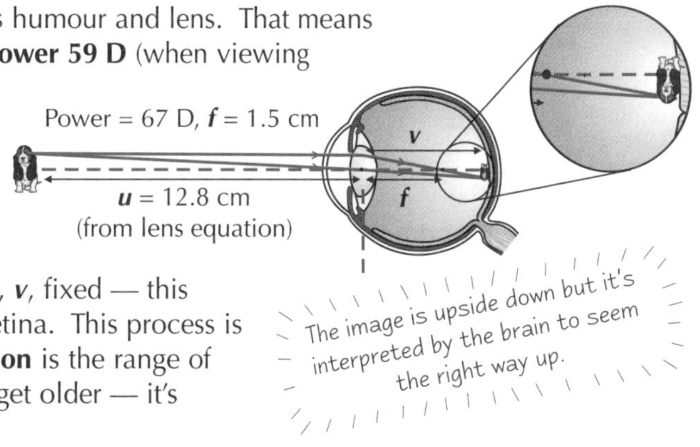

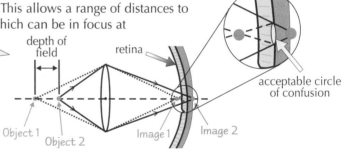

Power = 67 D, **f** = 1.5 cm

u = 12.8 cm
(from lens equation)

5) The eye's lens changes shape to keep the image distance, **v**, fixed — this means both near and far images can be focused at the retina. This process is called **accommodation**. The **amplitude of accommodation** is the range of powers that the eye can achieve. This decreases as you get older — it's about 11 D for children but almost 0 D for old people.

The image is upside down but it's interpreted by the brain to seem the right way up.

The Eye has a Limited **Depth of Field**

1) The **far point** is the **furthest distance** that the eye can focus comfortably. For normally sighted people that's **infinity**. When your eyes are focusing at the far point, they're **unaccommodated**.

2) The **near point** is the **closest distance** that the eye can focus on. For young people it's about 9 cm.

3) Points that are out of focus result in a **circle of confusion** on the retina. If the circle of confusion is **small** enough then the image is in focus. This allows a range of distances to be 'in focus' simultaneously. The range of distances which can be in focus at the same time is called the **depth of field**.

4) **Depth of field** is **greater** when the eye focuses on **further** distances. It also increases in **brighter light** when the pupil is smaller. In both cases, the rays of light entering the eye are at **shallower angles**, giving **smaller circles of confusion**.

depth of field

retina

acceptable circle of confusion

Object 1
Object 2
Image 1
Image 2

Practice Questions

Q1 Draw a ray diagram to show a young person's eye focusing at the near point. Assume that the eye's total power is 70 D. Mark on your diagram the distances **u**, **v** and **f**.

Q2 Sketch a diagram of the basic components of the eye and state the function of each one.

Q3 Write down an equation to show how **u**, **v** and **f** are related.

Q4 Define "depth of field". Explain why the depth of field increases with increasing light intensity.

Exam Questions

Q1 The power of an unaccommodated eye is 60 D.

(a) When the eye focuses at infinity, what will be the image distance, **v**? [2 marks]

(b) For the eye to focus on an object that is 30 cm away, what extra power must the lens produce? [3 marks]

Q2 A normally sighted eye has an unaccommodated power of 60 D and an amplitude of accommodation of 10 D. Find its near point and its far point. [6 marks]

The eyes are the window on the soul...

Or so they said in the 16th century. Sadly, philosophical wisecracks won't get you far with a question about amplitude of accommodation — better to write something about lenses changing shape and focal length decreasing.

Response of the Eye

These pages are for AQA A Option 6 and OCR A Option 02 only.

The retina contains thousands of light-sensitive cells, which can adapt readily to bright sunshine or near blackness.

The **Retina** has **Rods** and **Cones**

1) **Rods** and **cones** are cells at the back of the **retina** that respond to **light**. Light travels **through the retina** to the rods and cones at the back.

2) Rods and cones all contain chemical **pigments** that **bleach** when **light** falls on them. This bleaching stimulates the cell to send signals to the **brain** via the **optic nerve**.

3) The cells are **reset** (i.e. unbleached) by enzymes using **vitamin A** from the blood.

4) There's only **one** type of **rod** but there are **three** types of **cone**, which are sensitive to **red**, **green** and **blue** light.

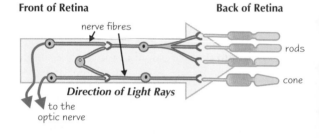

The **Cones** let you See in **Full Colour**

1) The red, green and blue **cones** each absorb a **range of wavelengths**.

2) The eye is **less responsive** to blue light than to red or green, so blues often look dimmer.

3) The brain receives signals from the three types of cone and interprets their **weighted relative strengths** as colour...

Example
Yellow light produces almost equal responses from the red and green cones.
Yellow light can therefore be 'faked' by combining red and green light of almost equal intensity — the electrical signal from the retina will be the same and the brain interprets it as 'yellow'.

4) **Any** colour can be produced by **combining** different intensities of **red**, **green** and **blue** light. Colour televisions work like this.

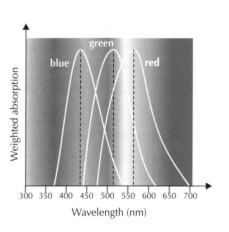

Colour has a **Psychological Impact**

1) The brain responds to different colours in different ways. **Brightly coloured** objects appear **larger** than duller ones and are more **attention grabbing**.

2) Colour also affects **mood** — for instance pale **green** tends to make people feel **calm**. Advertisers and architects take advantage of these different colour responses in their designs.

You Need Good **Spatial Resolution** to See **Details**   *AQA A Option 6 only*

1) Two objects can only be distinguished from each other if there's **at least one rod** or **cone between** the light from each of them. Otherwise the brain can't **resolve** the two objects and it 'sees' them as one.

2) **Spatial resolution** is **best** at the **yellow spot** — the **cones** are very **densely packed** here and each cone always has its **own nerve fibre**.

3) There are **no rods in the yellow spot**. This means that in **dim light**, when **cones don't work**, resolution is best slightly off the direct line of sight, where the **rods** are more **densely packed**. So if you want to watch something in the dark, don't look straight at it.

4) Away from the yellow spot, resolution is much worse. The light-sensitive cells are **not** as **densely packed** and the rods **share nerve fibres** — there are up to 600 rods per fibre at the edges of the retina.

Response of the Eye

The Eye Adapts to Varying Light Intensity

1) As **light** gets brighter or dimmer, the **pupil** can quickly **change size** to allow more or less light into the eye. The eye can also **adapt** in more subtle ways to **optimise vision**, using the different types of cell in the retina...

2) **Cones** need a **high light intensity** to respond and send signals to the brain, so they **don't work well** in **dim light**.

3) **Rods** respond in **dim light** — but in very bright light they are 'permanently' bleached.

LIGHT ADAPTION:

1) When light intensity **increases**, the **pigments** in the rods and cones **bleach**. If this happens **suddenly**, the cells become **saturated** and you're temporarily 'blinded'. It takes a **few seconds** for the eye to adjust by reducing the **sensitivity** of the cells. The eye is now **light-adapted**.

2) An eye adapted for **bright light** is predominantly using **cones**, which **don't share nerve cells**. You can see in **colour**, with increased resolving power for **detail** and **time** — which means you can see fast-moving objects clearly. This type of vision is called PHOTOPIC VISION.

DARK ADAPTION:

1) When light intensity **decreases** suddenly, the **cones** can't work, and you rely on **rods** to see. If the rods were previously saturated by bright light, they take about 40 minutes or so to regenerate their pigments and respond fully to light — when this happens the eye is **dark adapted**.

2) An eye adapted for **dim light** is predominantly using **rods**, which share nerve cells. The eye has **lower resolving power** so less detail can be seen, and it takes longer to process an image. In **very dim light**, you have **no colour** vision because the **cones don't work**. This type of vision is called SCOTOPIC VISION (or night vision).

Persistence of Vision means you Don't See Rapid Flickering
AQA A Option 6 only

1) **Nerve impulses** from the eye take about a fifth of a second to **decay**. So a very dim light flashing faster than **five times per second** (5 Hz) seems to be **on continuously**. This is called **flicker fusion**.

2) At **higher light intensity**, more nerves cells are 'firing' so a **higher frequency** is needed for flicker fusion to occur (about 70 Hz for very bright light).

3) Cinema and TV rely on **persistence of vision** to give the illusion of **smooth** rather than **jerky movements**.

Practice Questions

Q1 What are the main differences between rods and cones?

Q2 How does the eye adapt to bright light and what are the main features of photopic vision?

Q3 Explain what is meant by spatial resolution and how is it affected by the structure of the retina.

Q4 Sketch a graph showing how the cone cells in the retina respond to different wavelengths of light.

Exam Questions

Q1 a) Explain how the red, green and blue elements of a television can reproduce full colour. [4 marks]

    b) Explain how a television achieves the illusion of smooth movement. [3 marks]

Q2 a) What are the main features of scotopic vision? [4 marks]

    b) How do the eye's rod cells allow it to adapt to dim light? [5 marks]

I can see clearly now, the rain has gone...

*It probably **is** going to be a bright, bright, bright sunshiny day when you do your exam. But if you're really unlucky, you'll be crying with hayfever and you won't even be able to see the exam paper — so stock up on antihistamines now.*

Defects of Vision

These pages are for AQA A Option 6 and OCR A Option 02 only.

*Plenty of people don't have perfect vision, and need **auxiliary lenses** to correct their sight.*

Short Sight *is Corrected with* Diverging Lenses

1) **Short-sighted** (myopic) people are unable to focus on distant objects — this happens if their **far point** is **closer** than infinity (see p. 143).

2) Myopia occurs when the **cornea** and **lens** are too **powerful** or the **eyeball** is too **long**.

3) The focusing system is **too powerful** and images of distant objects are brought into focus in **front** of the retina.

4) A lens of **negative power** is needed to correct this defect — so a **diverging** lens is placed in front of the eye.

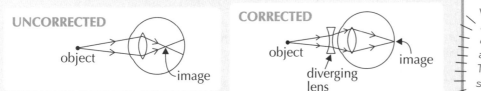

As well as correcting the far point, the diverging lens also makes the near point a little further away than it was. This isn't usually a problem — short-sighted people usually have a near point that is closer than normal anyway.

Long Sight *is corrected with* Converging Lenses

1) **Long-sighted** (hypermetropic) people are unable to focus clearly on near objects. This happens if their **near point** is **further** away than normal (25 cm or more).

2) Long sight occurs because the **cornea** and **lens** are too **weak** or the **eyeball** is too **short**.

3) The focusing system is **too weak** and images of near objects are brought to focus **behind** the retina.

4) A lens of **positive power** is needed to correct the defect — so a **converging** lens is placed in front of the eye.

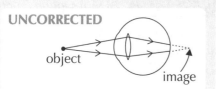

Long-sightedness is common among young children whose lenses have grown quicker than their eyeballs.

Older People *often have* Presbyopia   *OCR A Option 02 only*

1) As you get **older** your **lenses** lose their natural **elasticity** and can't become 'fat' enough for viewing close objects. Amplitude of **accommodation** falls from about 11 D in young children to close to **zero** for the **over 50s**.

2) As presbyopia develops, your **near point** gets gradually **further away**. The far point is unaffected.

3) This is **not** the same as **long sight** — people with presbyopia can be **short-sighted too**.

4) **Bifocal lenses** are used to correct the condition. Two lenses of **different powers** are arranged to correspond to the natural angles of viewing.

5) To view **distant** objects, you usually look **straight ahead** — so a lens to correct for short sight is needed at the **top** of a pair of bifocal glasses. To view **near** objects e.g. when reading, you're often looking **down** slightly — so a lens for near vision is needed at the **bottom**.

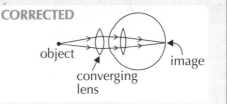

diverging lens

converging lens

Astigmatism *is Corrected with* Cylindrical Lenses

1) **Astigmatism** is caused by an irregularly shaped **cornea** which has **different focal lengths** for different **planes**. For instance, when **vertical lines** are in focus, **horizontal** lines might not be.

2) The condition is corrected with **cylindrical lenses**.

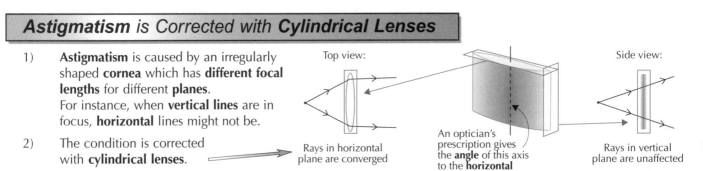

Top view:

Rays in horizontal plane are converged

An optician's prescription gives the **angle** of this axis to the **horizontal**

Side view:

Rays in vertical plane are unaffected

Defects of Vision

Choosing a **Lens** to Correct for **Short Sight** Depends on the **Far Point**

1) To correct for **short sight**, a **diverging** lens is chosen which has its **principal focus** at the eye's **faulty far point**.

2) The **principal focus** is the point that rays from a distant object **appear** to have come from (see p.142).

3) The lens must have a **negative focal length** which is the same as the **distance to the eye's far point**. This means that objects at **infinity**, which were out of focus, now seem to be in focus at the far point.

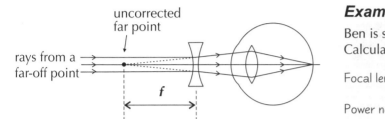

Example

Ben is short-sighted. His far point is 5 m.
Calculate the power of lens he needs to correct his vision.

Focal length, **f** = far point = −5 m

Power needed = $\dfrac{1}{f}$ = **−0.2 D** — The power's always **negative** to correct for **short** sight.

Calculations Involving **Long Sight** and **Presbyopia** Use the **Lens Equation**

1) People with these conditions have a near point which is too far away. An 'acceptable' near point is 25 cm.

2) To correct for **both conditions** a **converging lens** is used to produce a **virtual image** of objects 0.25 m away **at the eye's near point**. This means that close objects, which were out of focus, now seem to be in focus at the near point.

3) You can work out the **focal length**, and hence the **power** of lens needed, using the **lens equation** $\dfrac{1}{f} = \dfrac{1}{u} + \dfrac{1}{v}$

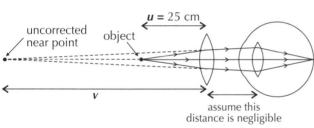

assume this distance is negligible

Example

Mavis can't read her book — her near point is 5 m.
What power of lenses does she need?

u = 0.25 m, **v** = −5 m (real is positive, virtual is negative)

$$\frac{1}{f} = \frac{1}{0.25} - \frac{1}{5}$$

$\dfrac{1}{f}$ = 3.8 so power = **+3.8 D** — The power's always **positive** to correct for **long** sight.

Practice Questions

Q1 Define the terms myopia, hypermetropia, presbyopia and astigmatism.

Q2 What type of auxiliary lenses are used to correct these conditions?

Q3 Define the terms near point and far point.

Q4 What is the difference between presbyopia and long-sightedness?

Exam Questions

Q1 A elderly man with short sight and presbyopia has a near point of 2 m and a far point of 4 m.

a) Calculate the power of auxiliary lens needed to correct the far point. [3 marks]

b) Calculate the power of lens required to correct the near point. [3 marks]

c) How is it possible to correct both problems in the same pair of spectacles? [1 mark]

Q2 a) What type of lenses are used to correct astigmatism? [1 mark]

b) Draw a diagram of a lens that would converge rays in the vertical plane but not in the horizontal plane. [2 marks]

You can't fly fighter planes if you wear glasses...

There's a hidden bonus to having dodgy eyes — in the exam, you can take your specs off (discreetly) and have a look at the lenses to remind yourself what type is needed to correct short sight, long sight, or whatever it is you have. Cunning.

The Physics of Hearing

These pages are for AQA A Option 6 *and* OCR A Option 02 *only.*

Ears are pretty amazing — they convert sound into electrical energy, using some tiny bones and lots of even tinier hairs.

Sound *Travels at* Different Speeds *in Different* Media   *AQA A Option 6 only*

When you're working out the **speed of sound**, *v*, there are different formulae for **fluids** (liquids or gases) and **solids**.

SOLIDS: $v = \sqrt{\dfrac{E}{\rho}}$

FLUIDS: $v = \sqrt{\dfrac{K}{\rho}}$

E and K are both **elastic moduli** (with units Pa or Nm⁻²). They can be thought of as a material's **resistance to change** in size or shape, i.e. its **stiffness**.
The values of **E** or **K** depend only on the **type of material** and can be looked up in a data book.

E = Young's modulus
K = bulk modulus
ρ is the density of the medium

1) In general, sound travels **faster** in **solids** than in liquids or gases.
2) In **fluids**, sound waves are always **longitudinal**.
3) In **solids**, sound waves can be **longitudinal or transverse**.
4) In **gases**, the speed of sound **increases** if the **temperature** increases.

SPEED OF SOUND IN DIFFERENT MEDIA:

Steel	5000 ms⁻¹
Water	1500 ms⁻¹
Air	340 ms⁻¹

The Intensity *of* Sound *is* Power *per* Unit Area

The **intensity** of a sound wave is defined as the amount of sound **energy** that passes **per second per unit area** (perpendicular to the direction of the wave).   That's **power per unit area**.

1) If the sound energy arriving at the ear per second is **P**, then the intensity of the sound is $I = \dfrac{P}{A}$

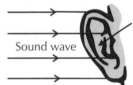

Area, **A**

Sound wave

The SI unit of **intensity** is Wm⁻², but you'll often see **decibels** used instead (see p.150).

2) For any wave, **intensity ∝ amplitude²** — so doubling the amplitude will result in four times the intensity.
3) Intensity is related to the **loudness** of sound (see p. 150).

Attenuation *is Progressive* Reduction *in Intensity*   *AQA A Option 6 only*

1) As a sound wave travels through a medium, the medium **absorbs** some of its **energy** — so the intensity **decreases**.
2) Intensity may also decrease if the wave **spreads out**, or through **diffraction** and **scattering**.
3) This **progressive reduction in intensity** is called **attenuation**, and is usually measured in **decibels per unit length**.

The Ear *has* Three Main Sections

The ear consists of three sections: the **outer** ear (**pinna** and **auditory canal**), the **middle ear** (**ossicles** and **Eustachian tube**) and the **inner ear** (**semicircular canals, cochlea** and **auditory nerve**).

1) The **tympanic membrane** (eardrum) separates the **outer** and **middle** ears.
2) Although separated, the **outer** and **middle** ears both contain **air** at **atmospheric pressure**, apart from slight pressure variations due to sound waves. This pressure is maintained by **yawning** and **swallowing** — the middle ear is opened up to the outside via the **Eustachian tube** (which is connected to the mouth).
3) The **oval** and **round windows** separate the **middle** and **inner** ears.
4) The **inner ear** is filled with fluid called **perilymph** (or **endolymph** in the **cochlear duct**). This fluid allows **vibrations** to pass to the basilar membrane in the **cochlea**.
5) The **semicircular canals** are involved with **maintaining balance**.

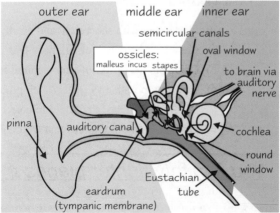

outer ear   middle ear   inner ear
semicircular canals
ossicles:
malleus incus stapes
oval window
to brain via auditory nerve
pinna
auditory canal
cochlea
round window
Eustachian tube
eardrum (tympanic membrane)

The Physics of Hearing

The *Ear* acts as a *Transducer*, converting *Sound Energy*...

1) The **pinna** (external ear) acts like a funnel, channelling sound waves into the auditory canal — this **concentrates** the energy onto a **smaller area** which increases the **intensity** slightly.

2) The sound wave consists of **variations** in **air pressure**, which make the **tympanic membrane** (eardrum) **vibrate**.

3) The tympanic membrane is connected to the **malleus** — one of the **three tiny bones** (**ossicles**) in the middle ear. The malleus then passes the **vibrations** of the eardrum on to the **incus** and the **stapes** (which is connected to the **oval window**).

4) As well as **transmitting vibrations**, the **ossicles** have **two** other functions — **amplifying** the sound signal and **reducing** the **energy reflected back** from the inner ear.

5) The **oval window** has a much **smaller area** than the **tympanic membrane**. Together with the **increased force** produced by the ossicles, this results in **greater pressure variations** at the oval window.

6) The **oval window** transmits vibrations to the **fluid** in the **inner ear**.

...into *Electrical Energy*

1) Pressure waves in the fluid of the **cochlea** make the **basilar membrane** vibrate. Different regions of this membrane have different **natural frequencies**, from 20 000 Hz near the middle ear to 20 Hz at the other end.

2) When a sound wave of a particular **frequency** enters the inner ear, one part of the basilar membrane **resonates** and so vibrates with a **large amplitude**.

3) **Hair cells** attached to the basilar membrane trigger **nerve impulses** at this point of greatest vibration.

4) These **electrical impulses** are sent, via the **auditory nerve**, to the **brain**, where they are interpreted as **sounds**.

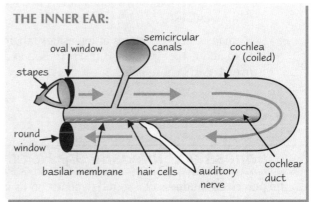

Practice Questions

Q1 What is the formula for the speed of sound in a gas?

Q2 Define intensity and attenuation.

Q3 Give three factors which cause attenuation of a sound wave.

Q4 Describe how a sound wave entering the ear results in a signal being transmitted along the auditory nerve.

Exam Questions

Q1 The bones of the middle ear can be modelled as a lever system, as shown in the diagram.

a) Explain how the ossicles act to increase the force at the oval window, F_2, compared to the force at the eardrum, F_1. [1 mark]

b) What other factor helps to increase the pressure at the oval window? [1 mark]

Q2 a) Sound waves travel at approximately 1500 ms⁻¹ in water. Calculate the elastic modulus of water. (Use ρ = 1000 kgm⁻³.) [2 marks]

b) The attenuation of sound waves in water is a lot less than in air. How does this affect communication by sound underwater? [2 marks]

Ears are like essays — they have a beginning, middle and end...

Or outer, middle and inner, if we're being technical. Learn what vibrates where, and you'll be fine.

Response of the Ear

These pages are for AQA A Option 6 and OCR A Option 02 only.

The ear's sensitivity depends on the **frequency** and **intensity** of sounds, and deteriorates as you get older.

Humans can Hear a Limited Range of Frequencies

1) Young people can hear frequencies ranging from about **20 Hz** (low pitch) up to **20,000 Hz** (high pitch). As you get older, the upper limit decreases.

2) Our ability to **discriminate between frequencies** depends on how **high** that frequency is. For example, between 60 and 1000 Hz, you can hear frequencies 3 Hz apart as **different pitches**. At **higher** frequencies, a **greater difference** is needed for frequencies to be distinguished. Above 10,000 Hz, pitch can hardly be discriminated at all.

3) The **loudness** of sound you hear depends on the **intensity** and **frequency** of the sound waves.

4) The **weakest intensity** you can hear — the **threshold of hearing**, I_0 — depends on the **frequency** of the sound wave.

5) The ear is **most sensitive** at around **3000 Hz**. For any given intensity, sounds of this frequency will be **loudest**.

6) Humans can hear sounds at intensities ranging from about 10^{-12} Wm^{-2} to 100 Wm^{-2}. Sounds **over 1 Wm^{-2}** cause **pain**.

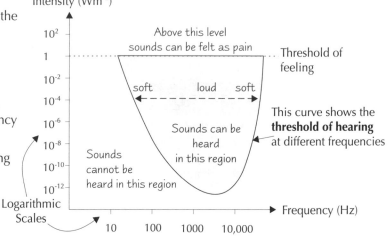

Loudness and Intensity are Related Logarithmically

The **perceived loudness** of a sound depends on its **intensity** (and its frequency — see above).

1) The relationship between **loudness** and **intensity** is **logarithmic**.

2) This means that **loudness**, **L**, goes up in **equal intervals** if **intensity**, **I**, increases by a **constant factor** (provided the frequency of the sound doesn't change).

3) E.g. if you **double** the intensity, **double it again** and so on, the **loudness** keeps going up in **equal steps**.

$$\Delta L \propto \log\left(\frac{I_2}{I_1}\right)$$

I_1 is the original intensity
I_2 is the new intensity
ΔL is increase in loudness

The Decibel Scale is used for Measuring Intensity Level

1) You can often measure loudness using a **decibel meter**. The **decibel scale** is a **logarithmic scale** which actually measures **intensity level**.

2) The **intensity level**, **IL**, of a sound of intensity **I** is defined as

$$IL = 10\log\left(\frac{I}{I_0}\right)$$

I = intensity
I_0 = threshold of hearing

3) I_0 is the **threshold of hearing** (the **lowest intensity** sound that can be heard) at a frequency of **1000 Hz**.

4) The value of I_0 is **1×10^{-12} Wm^{-2}**.

5) The units of **IL** are **decibels** (dB). Intensity level can be given in **bels** — one decibel is a tenth of a bel — but decibels are usually a more convenient size.

The dBA Scale is an Adjusted Decibel Scale   AQA A Option 6 only

1) The **perceived loudness** of a sound depends on its **frequency** as well as its intensity. Two different frequencies with the **same loudness** will have **different intensity levels** on the dB scale.

2) The **dBA** scale is an **adjusted decibel scale** which is designed to take into account the **ear's response** to **different frequencies**.

3) On the **dBA scale**, sounds of the **same intensity level** have the **same loudness** for the average human ear.

Response of the Ear

You can Generate *Curves of Equal Loudness*

1) Start by generating a **control frequency** of **1000 Hz** at a particular **intensity level**.

2) Generate another sound at a different frequency. Vary the volume of this sound until it appears to have the **same loudness** as the 1000 Hz frequency. Measure the **intensity level** at this volume.

3) Repeat this for several different frequencies, and plot the resulting curve on a graph.

4) Change the **intensity level** of the **control frequency** and repeat steps two and three.

5) If you measure **intensity level** in **decibels**, then the **loudness** of the sound is given in **phons**.

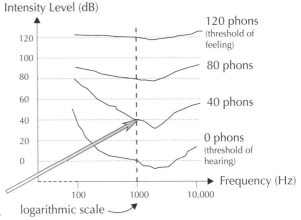

At 1000 Hz, the loudness in phons is the same value as the intensity level in decibels.

Hearing Deteriorates with Age and Exposure to Excessive Noise

1) As you get **older**, your hearing deteriorates **generally**, but **higher frequencies** are affected **most**.

2) Your ears can be damaged by **excessive noise**. This results in general hearing loss, but frequencies around **4000 Hz** are usually worst affected.

3) People who've worked with very **noisy machinery** have most hearing loss at the **particular frequencies** of the noise causing the damage.

4) **Equal loudness curves** can show hearing loss.

5) For a person with hearing loss, **higher intensity levels** are needed for the **same loudness**, when compared to a normal ear. A **peak** in the curve shows damage at a **particular** range of **frequencies**.

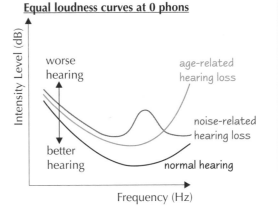

Equal loudness curves at 0 phons

Practice Questions

Q1 Define the threshold of hearing and sketch a graph that shows how it depends on frequency.
Q2 How are curves of equal loudness generated?
Q3 What is the dB scale? How is the dBA scale different?

Exam Questions

Q1 a) State the accepted value of the threshold of hearing, I_0, at 1000 Hz, in Wm^{-2}. [1 mark]
   b) What is the value of I_0 in decibels? [1 mark]

Q2 A small siren, which can be regarded as a point source, emits sound waves at a frequency of 3000 Hz. The intensity of the sound is 1 MWm^{-2} at a distance of 10 m away.
   a) Calculate the intensity level of the sound at 20 m away from the siren. Assume that the intensity of the sound is inversely proportional to the square of the distance from the siren. [3 marks]
   b) Why does the siren use a frequency of 3000 Hz? [1 mark]

Q3 The threshold of feeling is given as 120 dB. Calculate its intensity in Wm^{-2}. [3 marks]

Saved by the decibel....

It's medical fact that prolonged loud noise damages your hearing, so you should really demand ear protection before you agree to do the housework — some vacuum cleaners are louder than 85 dBA — the 'safe' limit for regular exposure.

The Heart

These pages are for AQA A Option 6 only.

There's a bit of biochemistry on this page — my, my, aren't you lucky people...

The **Heart** is a **Double Pump**

1) The heart is a **large muscle**. It acts as a **double pump**, with the **left**-hand side pumping blood from the **lungs** to the **rest of the body** and the **right**-hand side pumping blood from the **body** back to the **lungs**.

2) Traditionally, a diagram of the heart is drawn as though you're looking at it **from the front**, so the **right**-hand side of the heart is drawn on the **left**-hand side of the **diagram** and vice versa (just to confuse you).

3) Each side of the heart has **two chambers** — an **atrium** and a **ventricle** — separated by a **valve**.

4) **Blood** enters the **atria** from the veins, then the atria **contract**, squeezing blood into the **ventricles**. The **ventricles** then **contract**, squeezing the blood **out** of the heart into the **arteries**. The **valves** are there so that the blood doesn't go back into the atria when the ventricles contract.

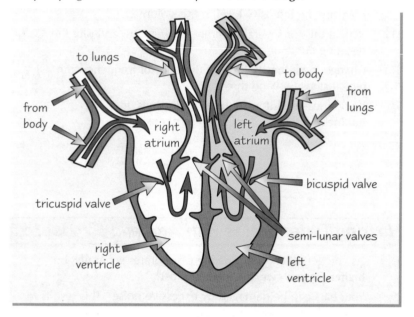

The Movement of **Sodium** and **Potassium Ions** Generates **Electrical Signals**

1) The movement of **sodium** and **potassium ions** across **cell membranes** in the heart is controlled by **diffusion** and a **sodium-potassium pump**.

2) The **pump** is a series of **chemical reactions** that move **sodium ions** (Na^+) in one direction through the membrane and **potassium ions** (K^+) in the other direction.

3) By **diffusion**, particles tend to move from **high** to **low concentrations** (along a **concentration gradient**). So, as the pump builds up the ion concentrations, the ions try to **diffuse back** in the **opposite** direction.

4) So far, so good — but here's where it gets a bit more complicated.

5) For most of the time, the membrane **only lets K^+ ions** diffuse through, and **not** the Na^+ ions. That means there's a **net flow** of **positive charge** in **one direction** and a **voltage** builds up until equilibrium is reached.

6) By convention, the potential on the **sodium side** of the membrane is taken to be **0 V** and so the membrane potential on the other side is **negative** (–80 mV for heart muscle). You've now got a **polarised** membrane.

7) If the membrane is **stimulated** by an **electrical signal** it suddenly becomes **permeable** to Na^+ as well. The sodium ions **rush** across the membrane towards the **negative side**, pushed by both the Na^+ **concentration gradient** and **electrostatic forces**.

8) This first **depolarises** the membrane, then **charges it up** the **other way** (called reverse-polarisation), reaching a **positive potential** of about 40 mV. This makes the heart muscle **CONTRACT**.

9) The membrane then becomes **impermeable** to Na^+ ions again, but **very permeable** to K^+ ions. The K^+ ions **diffuse** very quickly, repolarising the membrane, and the heart muscle **RELAXES**.

10) The Na-K pump then slowly restores the potential back to its **equilibrium polarised state**.

> The sudden flip in potential is called the **action potential**. When this happens at one part of a membrane, it triggers the part next to it to do the same and so an electrical signal passes down the membrane.

The Heart

The Heart's *Pacemaker* is the *Sinoatrial Node*

1) A group of specialised cells at the **sinoatrial (SA) node** (in the wall of the right atrium) produce **electrical signals** that pulse about **70 times a minute**.

2) These signals spread through the **atria** and make them **contract** via the **action potential** (see previous page).

3) The signals then pass to the **atrioventricular (AV) node**, which **delays** the pulse for about **0.1 seconds** before passing it on to the **ventricles**.

4) The ventricles **contract** and the process repeats.

The Heart can be *Monitored* by an *Electrocardiograph (ECG)*

1) The **potential difference** between the **polarised** and **depolarised** heart cells produces a **weak electrical signal** at the surface of the body. This is plotted against time to give an **electrocardiogram (ECG)**, which can provide useful information about the **condition** of the heart.

2) A **normal** ECG, covering a **single heartbeat**, has **three** separate parts: a **P** wave, a **QRS** wave and a **T** wave.

3) The **P wave** corresponds to the **depolarisation** and **contraction** of the **atria**.

4) The **QRS wave** (about 0.2 seconds later) corresponds to the **depolarisation** and **contraction** of the **ventricles**. This completely swamps the trace produced by the repolarisation and relaxation of the atria.

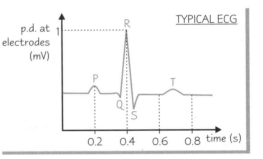

5) Finally, the **T wave** (another 0.2 seconds later) corresponds to the **repolarisation** and **relaxation** of the **ventricles**.

6) There are **twelve** standard ways of placing electrodes on the body to obtain an **ECG**, each producing a slightly different waveform. In all cases the signal is **heavily attenuated** (absorbed and weakened) by the body and needs to be amplified by a high impedance **amplifier**.

7) **Electrodes** are placed on the **chest** and the **limbs** where the arteries are close to the surface. The **right leg** is **never** used since it is **too far away** from the heart.

8) In order to get a **good electrical contact**, **hairs** and **dead skin** cells are removed and a **conductive gel** is used.

Practice Questions

Q1 Describe the basic structure of the heart and the passage of blood through the heart, lungs and the body.

Q2 Describe how a membrane in heart muscle is initially polarized and what happens when it is stimulated.

Q3 Sketch a typical ECG trace and indicate the main features.

Exam Questions

Q1 (a) What is an ECG? [1 mark]

(b) ECGs can be obtained by placing electrodes on different combinations of three limbs.
Why is the right leg never used? [1 mark]

(c) How is the patient's skin prepared so that a good quality ECG can be obtained? [2 marks]

Q2 (a) Explain the processes of depolarisation and repolarisation in the formation of an action potential. [5 marks]

(b) Briefly, explain how electrical signals are involved in the function of the heart. [4 marks]

Be still my beating sinoatrial node...

If you rely on the cast of ER to get your heart beating faster, console yourself that it's all very educational. Listen out for the machine that goes 'bip, bip, bip', and look for the P waves, QRS waves and T waves on the screen. If there aren't enough waves, the brave docs have to start shouting 'clear' and waving defibrillators around.

X-Ray Imaging

These pages are for AQA A Option 6 and OCR A Option 02 only.

X-rays are part of the spectrum of electromagnetic radiation. They're absorbed by bone but pass through soft tissue.

X-rays are Produced by Bombarding Tungsten with High Energy Electrons

1) In an X-ray tube, **electrons** are emitted from a **heated filament** and **accelerated** through a high **potential difference** (the **tube voltage**) towards a **tungsten anode**.

2) When the **electrons** smash into the **tungsten anode**, they **decelerate** and some of their **kinetic energy** is converted into **electromagnetic energy**, as **X-ray photons**. The tungsten anode emits a **continuous spectrum** of **X-ray radiation** — this is called **bremsstrahlung** ('braking radiation').

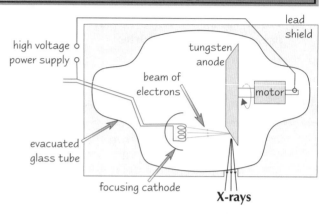

X-rays are also produced when beam electrons **knock out** other electrons from the **inner shells** of the **tungsten atoms**. Electrons in the atoms' **outer shells** move into the **vacancies** in the **lower energy levels**, and **release energy** in the form of **X-ray photons**.

3)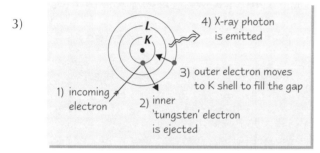

4) This process results in **line spectra** superimposed on a **continuous spectrum.**

5) Only about **1%** of the electrons' **kinetic energy** is converted into **X-rays**. The rest is converted into **heat**, so, to avoid overheating, the tungsten anode is **rotated** at about 3000 rpm. It's also **mounted** on **copper** — this **conducts** the heat away effectively.

Beam Intensity and Photon Energy can be Varied

The **intensity** of the X-ray beam is the **energy per second per unit area** passing through a surface (at right angles). There are two ways to increase the **intensity** of the X-ray beam:

1) Increase the **tube voltage**. This gives the electrons **more kinetic energy**. Higher energy electrons can **knock out** electrons from shells **deeper** within the tungsten atoms — giving more 'spikes' on the graphs. Individual **X-ray photons** also have **higher maximum energies**.
   Intensity is approximately **proportional** to **voltage squared**.

2) Increase the **current** supplied to the filament. This liberates **more electrons per second**, which then produce **more X-ray photons per second**. Individual **photons** have the **same energy** as before.
   Intensity is approximately **proportional** to **current**.
   The **intensity** of the X-ray beam is related to the **area under** the **graph**.

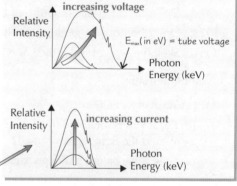

Radiographers try to Produce a Sharp Image and Minimise the Radiation Dose

1) Keep the photographic film **close to the patient** and make sure the **patient keeps still**.

2) Position the **X-ray tube further** from the **patient** (this means you need to **increase** the **exposure time**).

3) Use a **lead grid** to **prevent** scattered radiation from 'fogging' the film and **reducing contrast**.

4) Use an **intensifying screen**, which consists of crystals that **fluoresce** — they **absorb X-rays** and re-emit the energy as **visible light**, which helps to develop the photograph quickly. A shorter exposure time is needed, keeping the patient's radiation dose lower.

X-Ray Imaging

X-Rays are Attenuated when they Pass Through Matter

When X-rays pass through matter, they are **absorbed** and **scattered**. The intensity of the X-ray beam **decreases exponentially** with **distance from** the **surface**.

$$I = I_0 e^{-\mu x}$$

I is the intensity after a distance *x*, I_0 is the intensity at the surface and μ is the material's **attenuation coefficient** (in m^{-1})

Just for AQA A:

1) **Half-value thickness**, $x_{\frac{1}{2}}$, is the thickness of material required to **reduce** the **intensity** to **half** its **original value**.

This depends on the **attenuation coefficient** of the material, and is given by:

$$x_{\frac{1}{2}} = \frac{\ln 2}{\mu}$$

2) The **mass attenuation coefficient**, μ_m, for a material of density ρ is given by

$$\mu_m = \frac{\mu}{\rho}$$

X-rays are Absorbed More by Bone than Soft Tissue

1) A **photon** can be absorbed by an **electron**, which then **ejects** from its atom — causing the **photoelectric effect**.

2) For X-ray photons with energies of about **30 keV** this is the **main absorption process** for tissues in the body.

3) Materials with **higher atomic numbers** absorb **much more energy** — so tissues containing atoms with slightly **different atomic numbers** (e.g. **soft tissue** and **bone**) will produce a **large contrast** in the X-ray image.

4) Artificial **contrast media** are sometimes used if the tissues in the region of interest have very similar attenuation coefficients. E.g. the **barium meal** — **barium** has a **high atomic number**, so it shows up clearly in X-ray images. The barium meal is swallowed and can be followed as it moves along the patient's digestive tract.

Fluoroscopy and CT Scans use X-rays

1) **Moving images** can be created by **X-ray fluoroscopy**, using a **fluorescent screen** and an **image intensifier**. ⟹

2) **Computed tomography** (CT) scans produce an image of a **two-dimensional slice** through the body. An **X-ray beam rotates** around the body and is picked up by thousands of **detectors**. A computer works out how much attenuation has been caused by each part of the body and produces a very **high quality** image. However, the machines are **expensive**.

3) Both these techniques involve a **high radiation dose** for the patient.

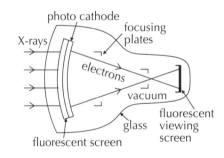

Practice Questions

Q1 Draw a diagram of an X-ray tube and explain how a typical X-ray spectrum is produced.

Q2 What measures can be taken to produce a high quality X-ray image while reducing the patient's radiation dose?

Exam Questions

Q1 An X-ray tube is connected to a potential difference of 30 kV.

a) Sketch a graph of relative intensity against photon energy (in eV) for the resulting X-ray spectrum, and indicate its main features. [3 marks]

b) Show how the graph in a) would change if the tube voltage were increased. [3 marks]

c) Why might a potential difference of 30 kV be used? [3 marks]

Q2 The half-value thickness for aluminium is 3 mm for 30 keV X-ray photons. What thickness of aluminium would reduce the intensity of a homogeneous beam of X-rays at 30 keV to 1% of its initial value? [4 marks]

I've got attenuation coefficient disorder — I get bored really easily...

X-ray images are just shadow pictures — bones absorb X-rays, stop them reaching the film and create a white 'shadow'.

Ultrasound Imaging

These pages are for AQA A Option 6 and OCR A Option 02 only.

Ultrasound is a 'sound' with higher frequencies than you can hear.

Ultrasound has a Higher Frequency than Humans can Hear

1) Ultrasound waves are **longitudinal** waves with **higher frequencies** than humans can hear (>20,000 Hz).

2) For **medical** purposes, frequencies are usually from **1** to **15 MHz**.

3) When an ultrasound wave meets a **boundary** between two **different materials**, some of it is **reflected** and some of it passes through (undergoing **refraction** if the **angle of incidence** is **not 90°**).

4) The **reflected waves** are detected by the **ultrasound scanner** and are used to **generate an image**.

The Amount of Reflection depends on the Change in Acoustic Impedance

1) The **acoustic impedance**, **Z**, of a medium is defined as:

   Z has units of $kgm^{-2}s^{-1}$.

ρ = density of the material, in kgm^{-3}
v = speed of sound in the medium, in ms^{-1}

2) If two materials have a **large difference** in **impedance**, then **most** of the energy is **reflected**. If the impedance of the two materials is the **same** then there is **no reflection**.

3) The **fraction** of wave **intensity** that has been reflected is called the **intensity reflection coefficient**, α.

$$\alpha = \frac{I_r}{I_i} = \left(\frac{Z_2 - Z_1}{Z_2 + Z_1}\right)^2$$

You don't need to <u>learn</u> this equation. Just practise using it.

There are Advantages and Disadvantages to Ultrasound Imaging

ADVANTAGES:

1) There are **no** known **hazards** — in particular, **no** exposure to **ionising radiation**.

2) It's good for imaging **soft tissues**, since you can obtain **real-time** images — X-ray fluoroscopy can achieve this, but involves a huge dose of radiation.

3) Ultrasound devices are relatively **cheap** and **portable**.

DISADVANTAGES:

1) Ultrasound **doesn't penetrate bone** — so **can't** be used to **detect fractures** or examine the **brain**.

2) Ultrasound **cannot** pass through **air spaces** in the body (due to the **mismatch in impedance**) — so can't produce images from behind the lungs.

3) **Resolution** is **poor** (about 10 times worse than X-rays), so you **can't see** fine **detail**.

Ultrasound Images are Produced Using the Piezoelectric Effect

1) **Piezoelectric crystals** produce a **potential difference** when they are **deformed** (squashed or stretched) — the rearrangement in structure displaces the **centres of symmetry** of their electric **charges**.

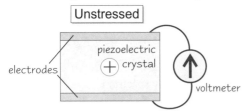
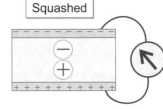
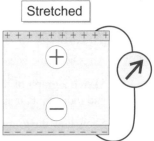

2) When you **apply a p.d.** across a piezoelectric crystal, the crystal **deforms**. If the p.d. is **alternating**, then the crystal **vibrates** at the **same frequency**.

3) A piezoelectric crystal can act as a **receiver** of ultrasound, converting **sound waves** into **alternating voltages**, and also as a **transmitter**, converting **alternating voltages** into **sound waves**.

4) Ultrasound devices use **lead zirconate titanate** (**PZT**) crystals. The **thickness** of the crystal is **half the wavelength** of the ultrasound that it produces. Ultrasound of this frequency will make the crystal **resonate** (like air in an open pipe — see p.26) and produce a large signal.

5) The PZT crystal is **heavily damped**, to produce **short pulses** and **increase** the **resolution** of the device.

Ultrasound Imaging

You need a **Coupling Medium** between the **Transducer** and the **Body**

1) **Soft tissue** has a very different **acoustic impedance** from **air**, so almost all the ultrasound **energy** is **reflected** from the surface of the body if there is air between the **transducer** and the **body**.

2) To avoid this, you need a **coupling medium** between the transducer and the body — this **displaces** the **air** and has an impedance much closer to that of body tissue.

3) The coupling medium is usually an **oil** or **gel** that is smeared onto the skin.

The rest of this page is for AQA A Option 6 only.

The **A-Scan** is a **Range Measuring** System

1) The **amplitude scan** (**A-Scan**) sends a short **pulse** of ultrasound into the body simultaneously with an **electron beam** sweeping across a cathode ray oscilloscope (**CRO**) screen.

2) The scanner receives **reflected** ultrasound pulses that appear as **vertical deflections** on the CRO screen.

3) **Weaker** pulses (that have travelled further in the body and arrive later) are **amplified** more to avoid the loss of valuable data — this process is called **time-gain compensation** (**TGC**).

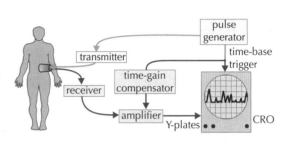

4) The **horizontal positions** of the reflected pulses indicate the **time** the 'echo' took to return, and are used to work out **distances** between structures in the body (e.g. the **diameter** of a **baby's head** in the uterus).

5) A **stream** of pulses can produce a **steady image** on the screen (due to **persistence of vision** — see p.145), although modern CROs can store a digital image after just one exposure.

In a **B-Scan**, the **Brightness** Varies

1) In a **brightness scan** (**B-Scan**), the electron beam sweeps **down** the screen rather than across.

2) The amplitude of the reflected pulses is displayed as the **brightness** of the spot.

3) You can use a **linear array** of transducers to produce a **two-dimensional** image.

Practice Questions

Q1   What are the main advantages and disadvantages of imaging using ultrasound?

Q2   How are ultrasound waves produced and received in an ultrasound transducer?

Q3   Define acoustic impedance.

Exam Questions

Q1 a) What fraction of intensity is reflected when ultrasound waves pass from air to soft tissue?
Use $Z_{air} = 0.430 \times 10^3$ kgm^{-2}s^{-1}, $Z_{tissue} = 1630 \times 10^3$ kgm^{-2}s^{-1}.   [2 marks]

   b) Calculate the ratio between the intensity of the ultrasound that **enters** the body when a coupling gel is used ($Z_{gel} = 1500 \times 10^3$ kgm^{-2}s^{-1}) and when none is used. Give your answer to the nearest power of ten.   [4 marks]

Q2 a) The acoustic impedance of a certain soft tissue is 1.63×10^6 kgm^{-2}s^{-1} and its density is 1.09×10^3 kgm^{-3}. Show that ultrasound travels with a velocity of 1.50 kms^{-1} in this medium.   [2 marks]

   b) The time base on a CRO was set to be 50 μscm^{-1}. Reflected pulses from either side of a foetal head are 2.4 cm apart on the screen. Calculate the diameter of the fetal head if the ultrasound travels at 1.5 kms^{-1}.   [4 marks]

Ultrasound — Mancunian for 'très bien'

You can use ultrasound to make images in cases where X-rays would do too much damage — like to check up on the development of a baby in the womb. You have to know what you're looking for though, or it just looks like a blob.

Magnetic Resonance Imaging

These pages are for OCR A Option 02 only.

This is all about absorption and re-emission of EM energy. I'm warning you now — this is not a fun page.

Atomic Nuclei can Behave like Magnets

1) **Protons** and **neutrons** possess a quantum property called **spin**, which makes them behave like **tiny magnets**.

2) If a nucleus has **even numbers** of **protons** and **neutrons** then the magnetic effects **cancel out**. A nucleus with an **odd number** of **protons** or **neutrons** has a **net spin** and is slightly **magnetic**.

3) The most important nucleus for **magnetic resonance imaging (MRI)** is **hydrogen**, which human bodies contain a lot of. A **hydrogen nucleus** has just **one proton**.

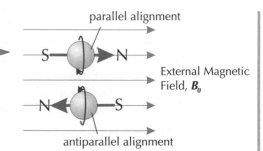

Protons Align themselves in a Magnetic Field

1) Normally, protons are orientated **randomly**, so their magnetic fields cancel out. When a strong **external magnetic field** is applied, the protons **align** themselves with the **magnetic field lines**.

2) Protons in **parallel alignment** point in the **same direction** as the external **magnetic field**. **Antiparallel alignment** means the protons point in the **opposite direction to the field**.

3) Protons align in these **two directions** in almost **equal numbers** (about 7 more per million are parallel).

4) antiparallel —N ⬤ S— E_2

$\Delta E \approx 2 \ \mu eV$

parallel —S ⬤ N— E_1

The **different alignments** correspond to **different energy levels**. The nuclei can **flip** between the two by **emitting** or **absorbing** a specific amount of **energy**.

5) The nuclei don't just stay still — they **precess** (wobble) **around** the **magnetic field lines** (like a **spinning top** precessing around gravitational field lines). They **don't** all precess **in phase** with each other.

6) All the protons **precess** at the **same frequency**, called the **Larmor frequency, f**. The value of f depends on the **strength** of the **magnetic field, B_0**.

$$f = \frac{\gamma B_0}{2\pi}$$

γ = gyromagnetic ratio (in HzT^{-1}),
B_0 = magnetic flux density of the external field (in T).

7) For protons, $\frac{\gamma}{2\pi} = 42.57$ $MHzT^{-1}$. For a typical B_0 of 1-2 T in an MRI scanner, f is in the **radio** frequency range.

Radio Waves can Make Protons Resonate

1) Protons have a **natural frequency** of oscillation in a magnetic field which is **equal** to the **Larmor frequency, f**. **Radio waves** at this frequency can make protons **resonate**.

2) The protons **absorb energy** from the radio waves and **flip** from **parallel** to **antiparallel alignment**.

3) Radio waves at the Larmor frequency also make the protons precess **in phase** with each other, producing a **rotating magnetic field** at **right angles** to the external field.

4) When the **external radio waves stop**, the protons return to their **original states** and **emit electromagnetic energy** as they do so — this emitted energy is the MRI signal.

5) The time taken for the protons to return to their original state is called the **relaxation time**, and is about **one second**.

6) **Relaxation times** depend on what molecules surround the protons, so they **vary** for **different tissue types**.

Magnetic Resonance Imaging

Magnetic Resonance can be used to Create Images

1) The patient lies in the centre of a huge **superconducting magnet**, which needs to be **cooled** by **liquid helium** — this is partly why the scanner is so **expensive**.

2) The magnet produces a **uniform magnetic field** of about **2 tesla**.

3) There are also three **gradient magnets** — electromagnets which produce a very small field (superimposed on the main one) that gradually **varies** from place to place. The **Larmor frequency** depends on the **total magnetic flux density**, so **different frequencies** of radio wave can be used to target **different sites** within the body.

4) Radio frequency **coils** are used to **transmit radio waves** and **receive** the MRI signal (the electromagnetic energy emitted by the relaxing protons). These coils are connected to a **computer**.

5) The computer **measures** various quantities of the MRI signal — amplitude, frequency, phase and relaxation times — and **analyses** them to generate an **image** of a **section** through the body.

Contrast can be Controlled by Varying the Pulses of Radio Waves

1) Radio waves are applied in **pulses**. Each short pulse **excites** the protons and then allows them to **relax** and emit a signal. The response of **different tissue types** can be enhanced by varying the **time between pulses**.

2) Tissues consisting of **large molecules** such as fat are best imaged using **rapidly repeated pulses**. This technique is used to image the internal **structure** of the body.

3) Allowing **more time** between each pulse enhances the response of **watery** substances. This method is used to image **diseased** areas.

MRI has Advantages and Disadvantages

ADVANTAGES:
1) There are **no** known **side effects**.
2) An image can be made for any slice in any **orientation** of the body.
3) High quality images can be obtained for **soft tissue** such as the **brain**.
4) **Contrast** can be **weighted** in order to investigate different situations.

DISADVANTAGES:
1) The imaging of bones is very poor.
2) Some people suffer from claustrophobia in the scanner.
3) Scans can be noisy (due to the switching of the gradient magnets) and take a long time.
4) MRI can't be used on people with pacemakers or surgical implants — the strong magnetic fields would be very harmful.
5) Scanners cost millions of pounds.

Practice Questions

Q1 Define the Larmor frequency and explain how radio waves at this frequency can cause resonance.

Q2 What is 'relaxation time' and how is this used to generate contrast within the images?

Q3 How do MRI scanners target protons from particular parts of the body?

Q4 Give two advantages of MRI compared to X-ray imaging.

Exam Questions

Q1 Draw a diagram to explain the meaning of the term 'precession'. [2 marks]

Q2 What causes protons in a magnetic field to 'flip' between parallel and antiparallel alignment? [2 marks]

Q3 Explain why hydrogen is the most important atom for MRI. [2 marks]

Precession — protons on parade

OK, so it hasn't been the easiest of pages. But at least now you know why people sit in vats of baked beans to raise money for their local hospital to buy an MRI scanner. Though perhaps you need A2 Psychology to understand the beans part.

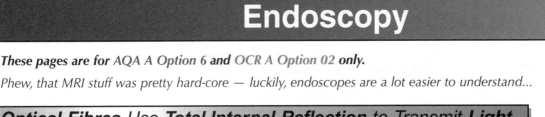

Endoscopy

These pages are for AQA A Option 6 and OCR A Option 02 only.

Phew, that MRI stuff was pretty hard-core — luckily, endoscopes are a lot easier to understand...

Optical Fibres Use Total Internal Reflection to Transmit Light

1) **Optical fibres** are a bit like electric wires — but instead of carrying current they **transmit light**.

2) A typical optical fibre consists of a **glass core** (about 5 μm to 50 μm in diameter) **surrounded** by a **cladding**, which has a slightly **lower refractive index**.

3) The **difference** in refractive index means that light travelling along the fibre will be **reflected** at the **cladding-core interface**.

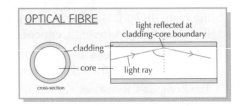

OPTICAL FIBRE

light reflected at cladding-core boundary
cladding
core
light ray
cross-section

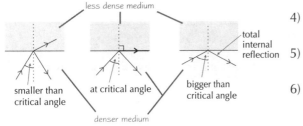

less dense medium

total internal reflection

smaller than critical angle | at critical angle | bigger than critical angle

denser medium

4) If the light ray's **angle of incidence** is **less than or equal** to a **critical angle**, some light will be **lost** out of the fibre.

5) But if the **angle of incidence** is **larger** than the **critical angle**, the light ray will be **completely reflected** inside the fibre.

6) This phenomenon is called **total internal reflection** and means that the ray **zigzags** its way along the fibre — so long as the fibre isn't too curved.

The Critical Angle for an Optical Fibre can be Worked Out

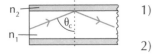

n_2
θ_c
n_1

1) The **critical angle**, θ_c, depends on the **refractive index** of the **core**, n_1, and **cladding**, n_2, in an optical fibre.

2) You can work out this value using the formula: $\longrightarrow$

$$\sin\theta_c = \frac{n_2}{n_1}$$

Example

An optical fibre consists of a core with a refractive index of 1.5 and cladding with a refractive index of 1.4.

a) What is the critical angle at the core-cladding boundary?

$$\theta_c = \sin^{-1}\left(\frac{n_2}{n_1}\right) = \sin^{-1}\left(\frac{1.4}{1.5}\right) = 69°$$

If some of this is sounding familiar, it's because you did it in AS. The equation for critical angle is derived from Snell's law of refraction.

b) Would total internal reflection occur if the incident angle of light is 70°?

70° > θ_c, so total internal reflection would occur.

Lots of Optical Fibres can be Bundled Together

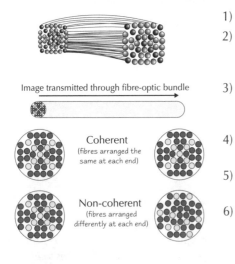

Image transmitted through fibre-optic bundle

Coherent
(fibres arranged the same at each end)

Non-coherent
(fibres arranged differently at each end)

1) An **image** can be transmitted along a **bundle** of optical fibres.

2) This can only happen if the **relative positions** of fibres in a bundle is the **same** between each end (otherwise the image would be jumbled up) — a fibre-optic bundle in this arrangement is said to be **coherent**.

3) The **resolution** (i.e. how much detail can be seen) depends on the **thickness** of the fibres. The thinner the fibres, the **more detail** that can be resolved — but thin fibres are more **expensive** to make.

4) Images can be **magnified** by making the diameters of the fibres get **gradually larger** along the length of the bundle.

5) If the relative **position** of the fibres **does not** remain the same between each end the bundle of fibres is said to be **non-coherent**.

6) **Non-coherent bundles** are much easier and **cheaper** to make. They **can't** transmit an **image** but they can be used to get **light** to hard-to-reach places — kind of like a flexible **torch**.

Endoscopy

Endoscopes Use Optical Fibres to Create an Image

1) An **endoscope** consists of a **long tube** containing **two bundles** of fibres — a **non-coherent** bundle to carry **light** to the area of interest and a **coherent** bundle to carry an **image** back to the eyepiece.

2) Endoscopes are widely used by surgeons to examine inside the body

3) An **objective lens** is placed at the **distal** end (**furthest from the eye**) of the **coherent** bundle to form an image, which is then transmitted by the fibres to the **proximal** end (**closest to the eye**) where it can be **viewed** through an **eyepiece**.

4) The **endoscope tube** can also contain a **water channel**, for cleaning the objective lens, a **tool aperture** to perform **keyhole surgery** and a **CO$_2$ channel** which allows CO$_2$ to be pumped into the area in front of the endoscope, making more room in the body.

Endoscopes are Used in Keyhole Surgery

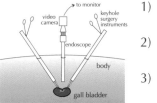

1) **Traditional** surgery needs a **large cut** to be made in the body so that there's **room** for the surgeons to get in and perform an **operation**.

2) This means that there's a **large risk of infection** to the exposed tissues and that permanent **damage** could be done to the patient's **body**.

3) New techniques in **minimally invasive surgery** (MIS or **keyhole surgery**) mean that only a **few small holes** need to be cut into the body.

4) An **endoscope** can be used in keyhole surgery to show the surgeon an **image** of the area of interest. **Surgical instruments** are passed through the endoscope tube, or through additional **small holes** in the body, so that the **operation** can be carried out.

5) **Common procedures** include the removal of the **gall bladder**, investigation of the **middle ear**, and removal of abnormal polyps in the **colon** so that they can be investigated for **cancer**.

6) **Recovery times** tend to be **quicker** for keyhole surgery, so the **patient** can usually **return home** on the **same day** — which makes it much **cheaper** for the hospital and **nicer** for the patient.

Practice Questions

Q1 What condition must be satisfied for total internal reflection to occur?

Q2 Explain the difference between a coherent and a non-coherent bundle of fibres.

Q3 What are the main features of an endoscope?

Q3 How have endoscopes revolutionised some surgical techniques?

Exam Questions

Q1 A beam of light is transmitted through an optical fibre with a refractive index of 1.35 for the core and 1.30 for the cladding.
 a) What is the critical angle for the core-cladding boundary? [1 mark]
 b) What will be observed if the beam of light is at an incident angle of 80°
  to the normal of the core-cladding boundary? [1 mark]

Q2 a) What would a coherent fibre-optic bundle be used for in an endoscope? [1 mark]
 b) Explain why there are lots of very thin fibres present in the coherent bundle. [2 marks]

If you ask me, physics is a whole bundle of non-coherentness...

If this is all getting too much, and your brain is as fried as a pork chipolata, just remember the wise words of revision wisdom from the great Spike Milligan — Ying tong, ying tong, ying tong, ying tong, ying tong, tiddly-aye-oh, tiddly-aye-oh...

Biological Effects of Radiation

This page is for OCR A Option 02 only.

Radiation — a thing that causes cancer — can actually be used to treat it. Another of life's little ironies.

Ionising Radiation Can Lead to Microscopic Damage...

1) **Ultraviolet**, **X-rays** and **γ-rays** from the electromagnetic spectrum and α and β **particles** from radioactive emissions are all classed as **ionising radiations**.

2) When they **interact** with matter they can **ionise atoms** to form **ions** — usually by **removing an electron** from the atom. This can cause **microscopic damage** to cells.

3) **DNA** in the cells can be **damaged** by either **direct** or **indirect action**:

Direct Action

1) **Direct action** involves the radiation directly damaging **DNA**, causing **mutations**.

2) The cell's ability to **reproduce** might be affected, making it **sterile** or leading to the growth of a **cancerous tumour**. The mutation might lead to **cell death** or, in certain circumstances, it might be **passed on** to **future generations** — which is a key feature of evolution.

Indirect Action

1) **Indirect action** (which is far more likely) involves radiation **splitting water molecules** into **highly reactive** particles called free radicals. These can in turn **damage DNA** — so causing **mutations**.

Microscopic Damage has Macroscopic Effects

1) **Ionising radiation** will cause **microscopic damage** to a cell. If the **DNA** is damaged, this could lead to **mutations**, like in cancerous tumours.

2) This **chance** of damaging the DNA is known as a **stochastic (random) effect**.

3) Stochastic effects will occur in cells up to a **threshold dose** of ionising radiation — **above** this threshold dose the cells will **die**.

4) Since **cell death** is **certain** above the threshold dose, the result is known as a **non-stochastic (not random) effect**.

5) If **enough cells** are **damaged** by ionising radiation, both stochastic and non-stochastic effects will be **visible** — if you can see the damage on a large scale it's called a **macroscopic effect**.

6) The **macroscopic effects** of radiation (i.e. the effects that are visible) include **tumours**, **skin burns**, **sterility**, **radiation sickness**, **hair loss** and **death** — nice.

Exposure, Absorbed Dose and Dose Equivalent

1) **Exposure** is defined as the **total charge produced** by ionising radiation **per unit mass of air**. The unit of exposure is the C kg^{-1}.

Air is used since its average atomic number is similar to that of body tissue.

2) The **absorbed dose** is more useful in medical physics because it is a measure of the **energy absorbed per unit mass**. The unit is the **gray (Gy)** where 1 Gy = 1 Jkg^{-1}.

3) The **dose equivalent, H,** takes into account the fact that the **amount of damage** to body tissue depends on the **type of ionising radiation**. The unit is the **sievert (Sv)** where 1 Sv = 1 Jkg^{-1}.

$$H = Q \times D$$

where H is the dose equivalent (Sv), Q is the quality factor, and D is the absorbed dose (Gy)

Radiation	Q
X-ray, β, γ	1
neutrons	5 - 20
α	20

4) **Each type** of radiation is assigned a weighted **quality factor, Q** — the **greater** the **damage** produced by a type of radiation, the **higher** the quality factor. ⟹

Example

Which absorbed dose would cause the greatest damage to a cell, 5 Gy of α or 10 Gy of β radiation?

To compare the potential damage to the cell you need to look at the dose equivalent of each type of radiation.

$H_\alpha = 20 \times 5 = 100$ Sv $H_\beta = 10 \times 1 = 10$ Sv

So 5 Gy of α radiation would cause more cell damage than 10 Gy of β.

Biological Effects of Radiation

This page is for AQA A Option 6 and OCR A Option 02 only.

Ionising Radiation Can Be Used to Treat Tumours
OCR A Option 02 only

1) **Cells** that are **rapidly multiplying** (e.g. cancerous **tumours**) are much **more likely** to be **damaged** by **radiation** than cells that are not — so, **carefully targeted** X-rays or γ radiation can **kill cancer cells** without doing too much damage to normal tissue (which isn't multiplying as quickly).

2) This kind of treatment is called **radiotherapy** and it is spread over many short sessions with **recovery time** in between.

3) The **ionising radiation** is aimed at the tumour from **different angles** so that the **tumour** is always **exposed** but normal tissue's exposure is limited.

Lasers Produce a Beam of Very Intense EM Radiation

1) **Lasers** produce beams of **EM radiation** with such **high intensities** that they can be used to **cut** through materials like a **scalpel**.

2) **Carbon dioxide lasers** produce **infrared light**, which not only **cuts** through body tissue, but also **heats** it up — this causes the **blood vessels** to shrink and harden, which **stops any bleeding**.

3) Lasers that produce **visible light** are used in conjunction with **endoscopes** (see p. 160) to **vaporise unwanted tissue** such as cancerous **tumours** and unsightly **birthmarks**.

4) **Laser eye surgery** is now becoming common practice, where the **shape** of the **cornea** can be altered by vaporising unwanted tissue, permanently **correcting eye defects**.

You Need to Be Careful How You Use Them Though...

1) The **eye** is particularly susceptible to **damage** from a laser beam because it **focuses** the light into an even smaller area — which makes the beam even **more intense**.

2) **Eye protection** must be worn at all times when working with lasers to avoid damage to the eye — people should **never look into the laser beam** and reflecting surfaces must be minimised to avoid **accidental exposure**.

3) Particular **care** must be taken with **infrared lasers** — **IR** beams are **invisible**, so exposure isn't noticed as quickly, and they can also **reflect** off surfaces that are not reflective to visible light.

Practice Questions

Q1 What is the difference between direct and indirect action, when describing the biological effects of ionising radiation?

Q2 What is the difference between stochastic and non-stochastic effects of radiation?

Q3 Define exposure, absorbed dose and dose equivalent. What are their units?

Q4 Explain why ionising radiation can be used to treat cancerous tumours.

Q5 Describe two different medical applications of lasers.

Exam Questions

Q1 A man of mass 70 kg accidentally swallows a source of α radiation (with a radiation quality factor of 20). The source emits 3×10^{10} particles per second (constant over this time period) and the energy of each particle is 8×10^{-13} J. The source was removed after 1000 seconds.

a) Assuming that all the ionising energy is distributed uniformly around the man's body, what is the absorbed dose that he receives? [2 marks]

b) Calculate the dose equivalent. [1 mark]

The biological effects of a page on radiation — a sore head...

Hoo-bleeding-rah — at last, you've made it to the end of a proper beast of a section/book/course. But before you shut this page and tootle off to do something more interesting, just make sure you know the difference between absorbed dose and equivalent dose and can work each one out — there's bound to be a question on them in your exam...

Answers

Section One — Dynamics
Page 3 — Momentum and Impulse

1) total momentum before collision = total momentum after [1 mark]
$(0.6 \times 5) + 0 = (0.6 \times -2.4) + 2v$
$3 + 1.44 = 2v$ [1 mark for working] $\Rightarrow v = 2.22 \ ms^{-1}$ [1 mark]

2) momentum before = momentum after [1 mark]
$(0.7 \times 0.3) + 0 = 1.1v$
$0.21 = 1.1v$ [1 mark for working] $\Rightarrow v = 0.19 \ ms^{-1}$ [1 mark]

Page 5 — Newton's Laws of Motion

1)

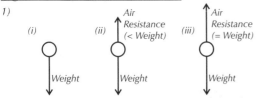

(i) Weight

(ii) Air Resistance (< Weight) Weight

(iii) Air Resistance (= Weight) Weight

[1 mark each diagram]

2)a) Force perpendicular to river flow = 500 – 100 = 400 N [1 mark]
Force parallel to river flow = 300 N
Resultant force = $\sqrt{400^2 + 300^2}$ = 500 N [1 mark]

b) $a = F/m$ [1 mark] = 500/250 = 2 ms^{-2} [1 mark]

Page 7 — Work and Power

1)a)

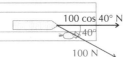

Force in direction of travel = $100 \ cos40° = 76.6 \ N$ [1 mark]
$W = Fs = 76.6 \times 1500 = 114 \ 900 \ J$ [1 mark]

b) Use $P = Fv$ [1 mark] = $100cos40° \times 0.8 = 61.3 \ W$ [1 mark]

2)a) Use $W = Fs$ [1 mark] = $20 \times 9.81 \times 3 = 588.6 \ J$ [1 mark]

b) Use $P = Fv$ [1 mark] = $20 \times 9.81 \times 0.25 = 49.05 \ W$ [1 mark]

Page 9 — Conservation of Energy

1)a) $KE = \frac{1}{2}mv^2$ and $PE = mgh$ [1 mark]. $\frac{1}{2}mv^2 = mgh \Rightarrow \frac{1}{2}v^2 = gh$
$\Rightarrow v^2 = 2gh = 2 \times 9.81 \times 2 = 39.24$ [1 mark]. $v = 6.26 \ ms^{-1}$ [1 mark]

b) 2 m — no friction means the kinetic energy will all change back into potential so he'll rise back to the same height as he started. [1 mark]

c) Put in some more energy by actively 'skating'. [1 mark]

2)a) The kinetic energy will be less after the collision. [1 mark]

b) total momentum before = total momentum after
$10 \ 000 \times 1 = 25 \ 000v$ [1 mark] so $v = 0.4 \ ms^{-1}$ [1 mark]

c) Before: $KE = 0.5 \times 10 \ 000 \times 1^2 = 5000 \ J$ [1 mark]
After: $KE = 0.5 \times 25 \ 000 \times 0.4^2 = 2000 \ J$ [1 mark]

Page 11 — Stretching Solids

1)a) Force is proportional to extension. The force is 1.5 times as great, so the extension will also be 1.5 times the original value.
Extension = $1.5 \times 4.0 \ mm = 6.0 \ mm$ [1 mark]

b) $F = ke \Rightarrow k = F/e$ [1 mark], $k = 10 \div 4.0 \times 10^{-3} = 2500 \ Nm^{-1}$ [1 mark]

c) One mark for any sensible point, e.g. The string now stretches much further for small increases in force. [1 mark]

2) The rubber band does not obey Hooke's law [1 mark] because when the force is doubled from 2.5 N to 5 N, the extension increases by a factor of 2.3 from 4.4 cm to 10.2 cm. [1 mark]

3)a) $Stress = \frac{force}{area} = \frac{300}{\pi r^2} = \frac{300}{\pi \left(\frac{1 \times 10^{-3}}{2} \right)^2} = 3.82 \times 10^8 \ Nm^{-2} \ or \ Pa$

[2 marks available — 1 mark for the correct equation including πr^2 for area, and 1 mark for the correct value for the stress]

b) Strain = extension/length = $4 \times 10^{-3}/2.00 = 2 \times 10^{-3}$ [1 mark]

c) $E_s = \frac{1}{2}F\Delta l = \frac{1}{2} \times 300 \times 4 \times 10^{-3} = 0.6 \ J$ [1 mark]

d) $E = \frac{stress}{strain} = \frac{3.82 \times 10^8}{2 \times 10^{-3}} = 1.91 \times 10^{11} \ Nm^{-2} \ or \ Pa$ [1 mark]

Section Two — Circular Motion & Oscillations
Page 13 — Circular Motion

1)a) $\omega = \frac{\theta}{t}$ [1 mark] so $\omega = \frac{2\pi}{3.2 \times 10^7} = 2.0 \times 10^{-7} \ rad \ s^{-1}$ [1 mark]

b) $v = r\omega$ [1 mark] = $1.5 \times 10^{11} \times 2.0 \times 10^{-7} = 30 \ kms^{-1}$ [1 mark]

c) $F = m\omega^2 r$ [1 mark] = $6.0 \times 10^{24} \times (2.0 \times 10^{-7})^2 \times 1.5 \times 10^{11}$
= $3.6 \times 10^{22} \ N$ [1 mark]

d) The gravitational force due to the Sun on the Earth [1 mark]

2)a) Gravity pulling down on the water gives a centripetal acceleration of 9.81 ms^{-2} [1 mark]. If the circular motion of the water needs a centripetal acceleration of less than 9.81 ms^{-2}, gravity will pull it in too tight a circle. The water will fall out of the bucket.

Since $a = \omega^2 r$, $\omega^2 = \frac{a}{r} = \frac{9.81}{1}$, so $\omega = 3.1 \ rad \ s^{-1}$ [1 mark]

$\omega = 2\pi f$, so $f = \frac{\omega}{2\pi} = 0.5 \ rev \ s^{-1}$ [1 mark]

b) Centripetal force = $m\omega^2 r = 10 \times 5^2 \times 1 = 250 \ N$ [1 mark].
This force is provided by both the tension in the rope, T, and gravity:
$T + (10 \times 9.81) = 250$. So $T = 250 - (10 \times 9.81) = 152 \ N$ [1 mark].

Page 15 — Simple Harmonic Motion

1)a) Simple harmonic motion is an oscillation in which an object always accelerates towards a fixed point [1 mark] with an acceleration directly proportional to its displacement [1 mark]. [The SHM equation would get you the marks if you defined all the variables.]

b) The acceleration of a falling bouncy ball is due to gravity. This acceleration is constant, so the motion is not SHM. [1 mark].

2)a) Maximum velocity = $(2\pi f)A = 2\pi \times 1.5 \times 0.05 = 0.47 \ ms^{-1}$ [1 mark].

b) Stopclock started when object released, so $x = A\cos(2\pi ft)$ [1 mark].
$x = 0.05 \times \cos(2\pi \times 1.5 \times 0.1) = 0.05 \times \cos(0.94) = 0.029 \ m$ [1 mark].

c) $x = A\cos(2\pi ft) \Rightarrow 0.01 = 0.05 \times \cos(2\pi \times 1.5t)$.
So $0.2 = \cos(9.4t) \Rightarrow \cos^{-1}(0.2) = 9.4t$. $9.4t = 1.37 \Rightarrow t = 0.15 \ s$.
[1 mark for working, 1 mark for correct answer]

Page 17 — Simple Harmonic Oscillators

1)a) Extension of spring = 0.20 – 0.10 = 0.10 m [1 mark]. Hooke's Law
gives $k = \frac{force}{extension}$, so $k = \frac{0.10 \times 9.8}{0.10} = 9.8 \ Nm^{-1}$ [1 mark].

b) $T = 2\pi \sqrt{\frac{m}{k}} \Rightarrow T = 2\pi \times \sqrt{\frac{0.10}{9.8}} = 2\pi \times \sqrt{0.01} = 0.63 \ s$ [1 mark].

c) $m \propto T^2$ so if T is doubled, T^2 is quadrupled and m is quadrupled [1 mark]. So mass needed = $4 \times 0.10 = 0.40 \ kg$ [1 mark].

2) $5T_{short \ pendulum} = 3T_{long \ pendulum}$ and $T = 2\pi \sqrt{\frac{l}{g}}$ [1 mark]. Let length of

long pendulum = l. So $5 \left(2\pi \sqrt{\frac{0.20}{g}} \right) = 3 \left(2\pi \sqrt{\frac{l}{g}} \right)$ [1 mark].

Dividing by 2π gives $5 \times \sqrt{\frac{0.20}{g}} = 3 \times \sqrt{\frac{l}{g}}$. Squaring and simplifying

gives $5 = 9l$ so length of long pendulum = 5/9 = 0.56 m [1 mark].

Page 19 — Free and Forced Vibrations

1)a) When a system is forced to vibrate at a frequency the same as its natural frequency [1 mark] it oscillates with a rapidly increasing amplitude [1 mark]. This is resonance.

b) See graph below. [1 mark] for showing a peak at the natural frequency, [1 mark] for sharp resonance.

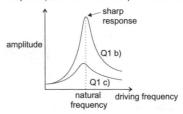

Answers

c) *See graph. [1 mark] for a smaller peak at the natural frequency [the natural frequency actually decreases a bit due to the damping, but you'll get the mark if the frequency stays the same in diagram].*

2)a) *A system is critically damped if it returns to rest in the shortest time possible [1 mark], when it's displaced from equilibrium and released.*

b) *e.g. suspension in a car [1 mark].*

Section Three — Waves
Page 21 — The Nature of Waves
1)a) *Use $v = \lambda f$ and $f = 1 / T$. So $v = \lambda / T \Rightarrow \lambda = vT$ [1 mark]*
$\lambda = 3\ ms^{-1} \times 6\ s = 18\ m$ *[1 mark]*
The vertical movement is irrelevant to this bit of the question.

b) *The trough to peak distance is twice the amplitude, so the amplitude is 0.6 m [1 mark].*

Page 23 — Electromagnetic Waves and Intensity
1) *At the same speed [1 mark]. Both are electromagnetic waves, so they travel at **c** in a vacuum [1 mark].*

2)a) *X-rays can be detected using photographic film [1 mark].*

b) *Medical X-rays [1 mark] rely on the fact that X-rays penetrate soft body tissue well but are blocked by bone [1 mark].* **OR** *Security scanners at airports [1 mark] rely on the fact that X-rays penetrate suitcases and clothes but are blocked by metal [1 mark].*

c) *The main difference between gamma rays and X-rays is that gamma rays arise from nuclear decay [1 mark] but X-rays are generated when metals are bombarded with electrons [1 mark].*

3) *Any of: unshielded microwaves, excess heat, damage to eyes from too bright light, skin cancer from UV, cancer or eye damage from UV, X-rays or gamma rays. [1 mark for the EM wave, 1 mark for the danger]*

Page 25 — Superposition and Coherence
1)a) *The frequencies and wavelengths of the two sources must be equal [1 mark] and the phase difference must be constant. [1 mark]*

b) *Interference will only be noticeable if the amplitudes of the two waves are of similar size. [1 mark]*

2)a) *180° (or $180° + 360n°$). [1 mark]*

b) *The displacements and velocities of the two points are equal in size [1 mark] but in opposite directions. [1 mark]*

Page 27 — Standing Waves
1)a)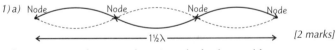
[2 marks]

b) *For a string vibrating at three times the fundamental frequency, length = $3\lambda / 2 \Rightarrow 1.2\ m = 3\lambda / 2 \Rightarrow \lambda = 0.8\ m$ [1 mark]*

c) *In a standing wave amplitude varies from a maximum at the antinodes to zero at the nodes. [1 mark] In an undamped progressive wave all the points have the same amplitude. [1 mark]*

d) *The displacements in successive antinodes are of equal size [1 mark] but opposite directions. [1 mark]*

Page 29 — Diffraction
1) *When a wavefront meets an obstacle, the waves will diffract round the corners of the obstacle. When the obstacle is much bigger than the wavelength, little diffraction occurs. In this case, the mountain is much bigger than the wavelength of short-wave radio. So the "shadow" where you cannot pick up short-wave is very long.*

[2 marks]

When the obstacle is comparable in size to the wavelength, as it is for the long-wave radio waves, more diffraction occurs. The wavefront re-forms after a shorter distance, leaving a shorter "shadow".

[2 marks]

2)a) *The pattern would get wider. [1 mark]*
*(Because $\sin \theta = \lambda / d$, and if **d** decreases θ increases.)*

b) $\sin \theta = \lambda / d$ *[1 mark]*
$\sin \theta = 6 \times 10^{-7}\ m / 0.1 \times 10^{-3}\ m$
$\sin \theta = 6 \times 10^{-3}$
$\theta = 0.344°$ *[1 mark]*

Page 31 — Two-Source Interference
1)a)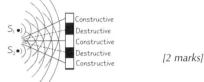
[2 marks]

b) *Light waves from separate sources are not coherent, as light is emitted in random bursts of energy. To get coherent light the two sets of waves must emerge from one source. [1 mark] A laser is used because it emits monochromatic coherent light [1 mark].*

2)a) $\lambda = v/f = 330 / 1320 = 0.25\ m$. *[1 mark]*

b) *Separation = $X = D\lambda/d$ [1 mark] = $(0.25 \times 7)/1.5 = 1.167\ m$ [1 mark].*

Page 33 — Diffraction Gratings
1)a) *Use $\sin \theta = n\lambda / d$*
For the first order, $n = 1$, so $\sin \theta = \lambda / d$ [1 mark]
No need to actually work out d. The number of lines per metre is 1 / d. So you can simply multiply the wavelength by that.
$\sin \theta = 600 \times 10^{-9} \times 4 \times 10^5 = 0.24 \Rightarrow \theta = 13.9°$ *[1 mark]*
For the second order, $n = 2$ and $\sin \theta = 2\lambda / d$. [1 mark]
$\sin \theta = 0.48 \Rightarrow \theta = 28.7°$ *[1 mark]*

b) *No. Putting $n = 5$ into the equation gives a value for $\sin \theta$ of 1.2, which is impossible. [1 mark]*

2) $\sin \theta = n\lambda / d$, *so for the first order maximum $\sin \theta = \lambda / d$ [1 mark]*
$\sin 14.2° = \lambda \times 3.7 \times 10^5 \Rightarrow \lambda = 663\ nm$ *[1 mark].*

Section Four — Gravitational & Electric Fields
Page 35 — Gravitational Fields
1) $g = \dfrac{GM}{r^2} \Rightarrow M = \dfrac{gr^2}{G} = \dfrac{9.81 \times (6400 \times 1000)^2}{6.67 \times 10^{-11}}$ *[1 mark]*
$= 6.02 \times 10^{24}\ kg$ *[1 mark]*

2)a) $g = \dfrac{GM}{r^2} = \dfrac{6.67 \times 10^{-11} \times 7.43 \times 10^{22}}{(1740 \times 1000)^2} = 1.64\ Nkg^{-1}$ *[1 mark]*

b) $E = -\dfrac{GMm}{r} = -\dfrac{6.67 \times 10^{-11} \times 7.43 \times 10^{22} \times 25}{(1740 + 10) \times 1000}$ *[1 mark]*
$= -7.08 \times 10^7\ J$ *[1 mark]*

Page 37 — Motion of Masses in Gravitational Fields
1)a) $T = \sqrt{\dfrac{4\pi^2 r^3}{GM}} = \sqrt{\dfrac{4\pi^2 \times [(6400 + 50) \times 1000]^3}{6.67 \times 10^{-11} \times 5.98 \times 10^{24}}}$ *[1 mark]*
$= 5154\ seconds\ OR\ 1.43\ hours$ *[1 mark]*

b) $v = \sqrt{\dfrac{GM}{r}} = \sqrt{\dfrac{6.67 \times 10^{-11} \times 5.98 \times 10^{24}}{(6400 + 40) \times 1000}}$ *[1 mark]*
$= 7.87\ kms^{-1}$

2) *Period = 24 hours = $24 \times 60 \times 60 = 86400\ s$ [1 mark]*
$T = \sqrt{\dfrac{4\pi^2 r^3}{GM}} \Rightarrow r = \sqrt[3]{\dfrac{T^2 GM}{4\pi^2}} = \sqrt[3]{\dfrac{86400^2 \times 6.67 \times 10^{-11} \times 5.98 \times 10^{24}}{4\pi^2}}$
$r = 4.23 \times 10^7\ m = 4.23 \times 10^4\ km$ *[1 mark]*
Height above Earth = $4.23 \times 10^4 - 6.4 \times 10^3 = 35900\ km$ [1 mark]

166

Answers

Page 39 — Electric Fields

1 (a) $E = V/d = 1500/(4.5 \times 10^{-3}) = 3.3 \times 10^5$ [1 mark] Vm^{-1} [1 mark]
   The field is perpendicular to the plates. [1 mark]
   (b) $d = 2 \times (4.5 \times 10^{-3}) = 9.0 \times 10^{-3}$ m [1 mark]
   $E = V/d \Rightarrow V = Ed$ $3.3 \times 10^5 \times 9 \times 10^{-3} = 3000$ V [1 mark]
2) Charge on alpha particle, $q = +2e = 2 \times 1.6 \times 10^{-19} = 3.2 \times 10^{-19}$ C
   Charge on gold nucleus, $Q = +79e = 79 \times 1.6 \times 10^{-19}$ C
   $= 1.3 \times 10^{-17}$ C [1 mark]

   $$F = \frac{kQq}{r^2} = \frac{9.0 \times 10^9 \times 3.2 \times 10^{-19} \times 1.3 \times 10^{-17}}{(5 \times 10^{-12})^2} \text{ [1 mark]}$$

   $= 0.0015$ N [1 mark] away from the gold nucleus [1 mark]

Page 41 — Gravitational and Electric Fields

1) $$F_g = \frac{Gm_1m_2}{r^2} = \frac{6.67 \times 10^{-11} \times (9.11 \times 10^{-31})^2}{(8 \times 10^{-10})^2} = 8.65 \times 10^{-53} \text{ N [1 mark]}$$

    $$F_e = \frac{kQ_1Q_2}{r^2} = \frac{9.0 \times 10^9 \times (1.60 \times 10^{-19})^2}{(8 \times 10^{-10})^2} = 3.60 \times 10^{-10} \text{ N [1 mark]}$$

    The electric force on each electron is much larger than the
    gravitational force, by a factor of over 10^{40} [1 mark].
2) a) Oil drop is stationary, so $mg = F_e = Vq/d$ [1 mark]
    $\Rightarrow q = mgd/V = (1.5 \times 10^{-14} \times 9.81 \times 0.03)/5000$ [1 mark]
    $= 8.8 \times 10^{-19}$ C [1 mark]
   b) The drop would accelerate towards the positive lower plate [1 mark].

Section Five — Capacitance
Page 43 — Capacitors

1) a) $C = 3(470 \times 10^{-6})$ [1 mark], so $C = 0.00141$ F or 1.41 mF [1 mark]

   b) $\frac{1}{C} = \frac{1}{470 \times 10^{-6}} + \frac{1}{470 \times 10^{-6}}$ [1 mark],
   so $C = 0.000235$ F or 235 μF [1 mark].

2) Capacitance $= \frac{Q}{V} =$ gradient of line $= \frac{660 \ \mu C}{3 \ V} = 220 \ \mu F$.

   [1 mark for 'gradient', 1 mark for correct answer.]
   Charge stored $= Q =$ area $= 15 \times 10^{-6} \times 66 = 990 \ \mu C$.
   [1 mark for 'area', 1 mark for correct answer.]

Page 45 — Energy Stored & Factors Affecting Capacitance

1) a) $E = \frac{1}{2}CV^2 = \frac{1}{2} \times 5 \times 10^{-4} \times 12^2 = 0.036$ J
   [1 mark for working, 1 mark for correct answer.]

   b) $E = \frac{1}{2}QV \Rightarrow Q = \frac{2E}{V} = \frac{2 \times 0.036}{12} = 0.006$ C
   [1 mark for working, 1 mark for correct answer.]
   You could have used $Q = CV$ here instead.

2) a) $C = \frac{\varepsilon_r \varepsilon_0 A}{d} = \frac{1.8 \times 8.85 \times 10^{-12} \times 0.3^2}{1 \times 10^{-3}} = 1.43 \times 10^{-9}$ F or 1.43 nF.
   [1 mark for working, 1 mark for correct answer.]

   b) The capacitance will be $\frac{1}{8}$ of its original value, since $\left(\frac{1}{2}\right)^2 \times \frac{1}{2} = \frac{1}{8}$
   [1 mark]. So $C = 1.8 \times 10^{-10}$ F or 0.18 nF [1 mark].

Page 47 — Charging and Discharging

1) a) The charge falls to 37% after RC seconds [1 mark],
   so $t = 1000 \times 2.5 \times 10^{-4} = 0.25$ seconds [1 mark]
   b) The charge falls to 50% after one half-life, $0.69RC$ seconds [1 mark],
   so $t = 0.69 \times 0.25 = 0.17$ seconds [1 mark]

   c) $Q = Q_0 e^{-\frac{t}{RC}}$ [1 mark], so after 0.7 seconds:

   $Q = Q_0 e^{\frac{0.7}{0.25}} = Q_0 \times 0.06$ [1 mark]. There is 6% of the initial charge
   left on the capacitor after 0.7 seconds [1 mark].
   d) i) The total charge stored will double [1 mark].
   ii) None [1 mark].   iii) None [1 mark].

Section Six — Electromagnetism
Page 49 — Magnetic Fields

1) a) $F = BIl = 2 \times 10^{-5} \times 3 \times 0.04$ [1 mark] $= 2.4 \times 10^{-6}$ N [1 mark]
   b) $F = BIl sin\theta = 2.4 \times 10^{-6} \times sin30° = 2.4 \times 10^{-6} \times 0.5$ [1 mark]
   $= 1.2 \times 10^{-6}$ N [1 mark]

Page 51 — Motion of Charged Particles in a B Field

1) a) $F = Bqv = 0.77 \times 1.6 \times 10^{-19} \times 5 \times 10^6$ [1 mark]
   $= 6.16 \times 10^{-13}$ N [1 mark]
   b) The force acting on the electron is always at right angles to its
   velocity. This is the condition for circular motion. [1 mark]
2) Electromagnetic force = Centripetal force [1 mark]
   so, $Bqv = mv^2 / r$ [1 mark]

   so, $r = \frac{mv}{Bq} = \frac{9.11 \times 10^{-31} \times 2.3 \times 10^7}{0.6 \times 10^{-3} \times 1.6 \times 10^{-19}} = 0.218$ m [1 mark]

3) The charged particle will be undeflected if the forces are balanced.
   This happens when $v = E / B$. [1 mark]

   so, $v = \frac{3.75 \times 10^4}{1.5 \times 10^{-3}} = 2.5 \times 10^7 ms^{-1}$ [1 mark]

Page 53 — Electromagnetic Induction

1) a) $\phi = BA$ [1 mark] $= 2 \times 10^{-3} \times 0.23 = 4.6 \times 10^{-4}$ Wb [1 mark]
   b) $\Phi = BAN$ [1 mark] $= 2 \times 10^{-3} \times 0.23 \times 150 = 0.069$ Wb [1 mark]
   c) $V = \frac{d\Phi}{dt}$

   $= \frac{(B_{start} - B_{end})AN}{t}$

   $= \frac{(2 \times 10^{-3} - 1.5 \times 10^{-3})(0.23 \times 150)}{2.5} = 6.9 \times 10^{-3}$ V

   [3 marks available, one for each stage of the workings]
2) a) $V = Blv$ [1 mark] $= 60 \times 10^{-6} \times 30 \times 100 = 0.18$ V [1 mark]
   b) [1 mark]

Page 55 — Transformers and Alternators

1) a) $\frac{V_p}{V_s} = \frac{N_p}{N_s}$ [1 mark] so, $N_s = \frac{45 \times 150}{9} = 750$ turns [1 mark]

   b) $\frac{V_p}{V_s} = \frac{I_s}{I_p}$ [1 mark] so, $I_s = \frac{V_p I_p}{V_s} = \frac{9 \times 1.5}{45} = 0.3$ A [1 mark]

   c) efficiency $= \frac{V_s I_s}{V_p I_p}$ [1 mark] $= \frac{10.8}{9 \times 1.5} = 0.8$ (i.e. 80%) [1 mark]

2) a) $\Phi = BANsin\theta$ [1 mark]
   $= 0.9 \times 0.01 \times 500 \times sin30° = 2.25$ Wb [1 mark]
   b) Peak e.m.f. when $cos\theta = \pm1$, giving: $V = \pm BAN2\pi f$ [1 mark]
   So, peak e.m.f. is: $V = \pm 0.9 \times 0.01 \times 500 \times 2 \times \pi \times 20$ [1 mark]
   $V = \pm 565.5$ V [1 mark]

Section Seven — Quantum Phenomena
Page 57 — Charge/Mass Ratio of the Electron

1 a) 1000 eV [1 mark]
   b) 1000 eV $\times 1.6 \times 10^{-19}$ J/eV $= 1.6 \times 10^{-16}$ J [1 mark]
   c) Kinetic energy $= \frac{1}{2}mv^2 = 1.6 \times 10^{-16}$ J [1 mark]
   $v^2 = (2 \times 1.6 \times 10^{-16}) \div (9.1 \times 10^{-31}) = 3.5 \times 10^{14}$
   $v = 1.9 \times 10^7 ms^{-1}$ [1 mark]
   Divide by 3.0×10^8: 6.3% of the speed of light [1 mark]

Answers

2 *[Your answer will depend on which experiment you describe, e.g.]
Electrons are accelerated using an electron gun [1 mark]. A magnetic
field [1 mark] exerts a centripetal force [1 mark] on the electrons
making them trace a circular path. By measuring the radius of this
path and equating the magnetic and centripetal forces [1 mark] you
can calculate e/m.
[1 mark for quality of written communication]
"You may be awarded a mark..." means that you almost certainly will.*

Page 59 — Millikan's Oil-Drop Experiment

1 a) *The diagram should show two arrows of equal length [1 mark] with
weight acting downwards and electrical force upwards [1 mark].*

   b) $mg = QV/d$ *[1 mark]*
   $Q = mgd/V = 1.63 \times 10^{-14}\ kg \times 9.81 \times 3.00 \times 10^{-2}\ m/5000$ *[1 mark]*
   $= 9.6 \times 10^{-19}\ C$ *[1 mark]*

   c) *6 (since value is $6 \times 1.6 \times 10^{-19}$ C) [1 mark]*

Page 61 — Light — Newton vs Huygens

1 *Light consists of particles [1 mark]. The theory was based on the laws
of motion with the straight-line motion of light as evidence [1 mark].*

2 *Most scientists in the 18th century supported Newton's corpuscular
theory [1 mark]. He said that light was made up of particles that
obey his laws of motion [1 mark]. In Huygens' wave theory, light is a
wave [1 mark]. This is supported by diffraction and interference in
Young's double-slit experiment [1 mark].
Maxwell described light as an electromagnetic wave [1 mark]
consisting of oscillating electric and magnetic fields.*

Page 63 — The Photoelectric Effect

1 *The plate becomes positively charged [1 mark]. Electrons in the
metal absorb energy from the UV light and leave the surface [1 mark].*

2 a) $\Phi = hf - (\frac{1}{2}mv^2)_{max} = hc/\lambda - (\frac{1}{2}mv^2)_{max}$ *[1 mark]*
   $= (6.6 \times 10^{-34} \times 3.0 \times 10^8)/(0.5 \times 10^{-6}) - 2.0 \times 10^{-19}\ J$ *[1 mark]*
   $= 1.96 \times 10^{-19}\ J$ *[1 mark]*

   b) $hc/\lambda < 1.96 \times 10^{-19}\ J$ *[1 mark]*
   $\lambda > (6.6 \times 10^{-34} \times 3.0 \times 10^8)/1.96 \times 10^{-19} = 1.01\ \mu m$ *[1 mark]*

3 $\Phi = 2.2\ eV = 2.2 \times 1.6 \times 10^{-19}\ J = 3.52 \times 10^{-19}\ J$
   $(\frac{1}{2}mv^2)_{max} = hf - \Phi = hc/\lambda - \Phi$ *[1 mark]*
   $= [6.6 \times 10^{-34} \times (3.0 \times 10^8/350 \times 10^{-9})] - 3.52 \times 10^{-19}$
   $= 2.14 \times 10^{-19}\ J$ *[1 mark]*
   Stopping potential, $V_S = (\frac{1}{2}mv^2)_{max}/e$ *[1 mark]*
   $\Rightarrow V_S = 2.14 \times 10^{-19}/1.6 \times 10^{-19}\ C = 1.3\ V$ *[1 mark]*

Page 65 — Wave-Particle Duality

1 a) i) *Velocity is given by* $\frac{1}{2}mv^2 = eV$ *[1 mark]*
   $\Rightarrow v^2 = 2eV/m \Rightarrow v = 1.3 \times 10^7\ ms^{-1}$ *[1 mark]*
   ii) *de Broglie* $\lambda = h/mv$ *[1 mark]* $\Rightarrow \lambda = 5.5 \times 10^{-11}\ m$ *[1 mark]*

   b) *This is in the X-ray region of the EM spectrum [1 mark].*

2   $E_k = \frac{1}{2}mv^2 \Rightarrow v = \sqrt{\frac{2E_k}{m}}$ *and* $\lambda = \frac{h}{mv}$ *[1 mark]*

   Substitute for v, $\lambda = \dfrac{h}{m\sqrt{\frac{2E_k}{m}}} = \dfrac{h}{\sqrt{\frac{2E_k m^2}{m}}}$ *[1 mark]*

   So, $\lambda = \dfrac{h}{\sqrt{2E_k m}}$ *[1 mark]*

Page 67 — Electron Energy Levels

1 a) $3.8 \times 10^{-5}\ eV = 3.8 \times 10^{-5} \times 1.6 \times 10^{-19} = 6.1 \times 10^{-24}\ J$
   $\Delta E = hf \Rightarrow f = \Delta E/h = 6.1 \times 10^{-24}/6.6 \times 10^{-34}$ *[1 mark]*
   $= 9.2 \times 10^9\ Hz$ *[1 mark]*

   b) 9.2×10^9 *oscillations occur every second [1 mark].*

2 *[Maximum 5 marks, 1 mark each for any of the following]*
   – Electrons are "pumped" to a high energy state
   – Electrons fall down to a metastable state spontaneously
   *– Some electrons fall to the ground state, giving out photons of a
   specific frequency*
   – This stimulates the emission of other photons of same frequency
   – All photons emitted are in phase
   – Mirrors cause further stimulation and amplification

Section Eight — Thermal Physics & Kinetic Theory

Page 69 — Ideal Gases

1) a) *Number of moles* $= \dfrac{mass\ of\ gas}{molar\ mass} = \dfrac{0.014}{0.028} = 0.5$ *[1 mark]*

   b) $pV = nRT$, *so* $p = \dfrac{nRT}{V}$ *[1 mark]*

   $p = \dfrac{0.5 \times 8.31 \times 300}{0.01} = 125\,000\ Pa$ *[1 mark]*

2) a) $\dfrac{pV}{T} = constant$

   At ground level, $\dfrac{pV}{T} = \dfrac{1 \times 10^5 \times 10}{293} = 3410\ JK^{-1}$ *[1 mark]*
   Higher up, pV/T will equal this same constant. *[1 mark]*
   So higher up, $p = \dfrac{constant \times T}{V} = \dfrac{3410 \times 260}{25}$

   $p = 35\,500\ Pa$ *[1 mark]*

   b) *Any two reasonable assumptions
   e.g. no helium gas is lost from the balloon — the mass stays fixed,
   the helium behaves like an ideal gas (or a suitable property of ideal
   gases is given). [2 marks — 1 for each]*

Page 71 — Molecular Kinetic Theory & Internal Energy

1) a) *Mass of 1 molecule* $= \dfrac{mass\ of\ 1\ mole}{N_A} = \dfrac{2.8 \times 10^{-2}}{6.02 \times 10^{23}} = 4.65 \times 10^{-26}\ kg$

   [1 mark]

   b) $\frac{1}{2}m\overline{c^2} = \frac{3kT}{2}$ *Rearranging gives:*

   $\overline{c^2} = \dfrac{3kT}{m} = \dfrac{3 \times 1.38 \times 10^{-23} \times 300}{4.65 \times 10^{-26}} = 2.67 \times 10^5\ m^2s^{-2}$ *[2 marks]*

   Typical speed = r.m.s. speed $= \sqrt{2.67 \times 10^5} = 517\ ms^{-1}$ *[1 mark]*

2) a) *Speed* $= \dfrac{distance}{time}$ *so time* $= \dfrac{distance}{speed}$

   The time $= \dfrac{8.0\ m}{400\ ms^{-1}} = 0.02\ s$ *[1 mark]*

   b) *Although the particles are moving at $400\ ms^{-1}$, they are frequently
   colliding with other particles. [1 mark]
   This means their forward motion is limited and so they only slowly
   move from one end of the room to the other. [1 mark]*

   c) *At 30 °C the average speed of the particles will be slightly faster
   [1 mark] since the absolute temperature has risen from 293 K to
   303 K and the temperature determines the average speed. [1 mark]*

Page 73 — Specific Heat Capacity & Specific Latent Heat

1) *Electrical energy supplied:*
   $\Delta Q = VI\Delta t = 12 \times 7.5 \times 180 = 16200\ J$ *[1 mark]*
   The temperature rise is $12.7 - 4.5 = 8.2\ °C$

   Specific heat capacity: $c = \dfrac{\Delta Q}{m\Delta\theta}$ *[1 mark]*

   $= \dfrac{16200}{2 \times 8.2} = 988\ Jkg^{-1}°C^{-1}$ *[1 mark]*

   You need the right unit for the third mark — J kg^{-1} K^{-1} would be right too.

2) *Total amount of energy needed to boil all the water:*
   $\Delta Q = l\Delta m = 2.26 \times 10^6 \times 0.5 = 1.13 \times 10^6\ J$ *[1 mark]*
   3 kW means you get 3000 J in a second, so
   time in seconds $= 1.13 \times 10^6 / 3000$ *[1 mark]* $= 377\ s$ *[1 mark]*

Page 75 — Thermodynamics

1) a) $Q = \Delta U - W$ *[1 mark]*
   *where Q is the heat energy supplied to the system, ΔU is the internal
   energy gained by the system and W is the work done on the system.
   [2 marks for all three correct, lose a mark for each mistake.]*

   b) $Q = \Delta U - W \Rightarrow \Delta U = Q + W$
   $\Delta U = -20 + 60$ *[1 mark]* $= 40\ J$ *[1 mark]*

   c) *The temperature will rise during this change [1 mark].*

Page 77 — The Boltzmann Factor

1) a) $kT = 1.38 \times 10^{-23} \times 300 = 4.1 \times 10^{-21}$ J [1 mark]
   b) For two bonds, $\varepsilon = 2 \times 3.2 \times 10^{-20} = 6.4 \times 10^{-20}$ J [1 mark]
   c) $\dfrac{\varepsilon}{kT} = \dfrac{6.4 \times 10^{-20}}{4.1 \times 10^{-21}} \approx 15$ [1 mark]
   d) With an ε/kT ratio of about 15, processes can take place using random thermal energy [1 mark]. Although the average energy of a water particle is much less than it needs to escape [1 mark] some particles will have enough energy to break their bonds and escape, meaning that the tank must be topped up to replace the water lost by evaporation [1 mark].

Section Nine — Nuclear and Particle Physics
Page 79 — Probing to Detemine Structure

1) a) The majority of alpha particles are not scattered because the nucleus is a very small part of the whole atom and so the probability of getting near it is small [1 mark]. Most alpha particles pass undeflected through the empty space around the nucleus [1 mark].
   b) Alpha particles and atomic nuclei are both positively charged [1 mark]. If an alpha particle travels close to a nucleus, there will be a significant electrostatic force of repulsion between them [1 mark]. This force causes the alpha particle to be deflected from its original path. [1 mark]
2) a) All particles have wave-like properties, with an associated wavelength [1 mark]. If the wavelength of a beam of particles is similar to the atomic spacing of the material it's passing through, the beam will produce a diffraction pattern. [1 mark]
   b) Electrons are not affected by the strong nuclear force. [1 mark]
   c) Maximum diffraction occurs when the nucleus is the same size as the wavelength of the electrons [1 mark]. Larger nuclei cause less diffraction for the same electron energy [1 mark].

Page 81 — Nuclear Radius and Density

1) a) Rearrange $r = r_0 A^{1/3}$ in terms of r_0 [1 mark], then substitute the given values of r and A:
   $r_0 = \dfrac{r}{A^{1/3}} = \dfrac{3.2 \times 10^{-15}}{12^{1/3}} = 1.4 \times 10^{-15}$ m [1 mark]
   b) For a radium nucleus with $A = 226$:
   $r = r_0 A^{1/3} = 1.4 \times 10^{-15} \times 226^{1/3} = 8.53 \times 10^{-15}$ m [1 mark]
   c) Mass $= 226 \times 1.66 \times 10^{-27}$ kg $= 3.75 \times 10^{-25}$ kg [1 mark]
   Volume $= \dfrac{4}{3}\pi r^3 = \dfrac{4}{3}\pi \left(8.53 \times 10^{-15}\right)^3 = 2.6 \times 10^{-42}$ m³ [1 mark]
   So density $(\rho) = \dfrac{m}{v} = \dfrac{3.75 \times 10^{-25}}{2.6 \times 10^{-42}} = 1.44 \times 10^{17}$ kgm⁻³ [1 mark]
2) The mass density of a gold nucleus is much larger than the mass density of a gold atom [1 mark]. This implies that the majority of a gold atom's mass is contained in the nucleus [1 mark]. The nucleus is small compared to the size of the atom [1 mark]. There must be a lot of nearly empty space inside each atom [1 mark].

Page 83 — The Strong Interaction

1) a) $F = \dfrac{1}{4\pi\varepsilon_o}\dfrac{Q_1 Q_2}{r^2} = \dfrac{1}{4\pi\left(8.85 \times 10^{-12}\right)}\dfrac{\left(1.6 \times 10^{-19}\right)\left(1.6 \times 10^{-19}\right)}{\left(9 \times 10^{-15}\right)^2} = 2.8N$
   [1 mark for working, 1 mark for correct answer]
   b) The electrostatic force will increase [1 mark].
   c) There is no electrostatic force between a proton and a neutron [1 mark] because a neutron has no charge [1 mark].
2) a) The strong interaction must be repulsive at very small nucleon separations to prevent the nucleus being crushed to a point [1 mark].
   b) Beyond 10 fm, the strong interaction is smaller than the electrostatic force [1 mark]. This means the protons in the nucleus would be forced apart. So a nucleus bigger than this would be unstable. [1 mark]

Page 85 — Radioactive Emissions

1)
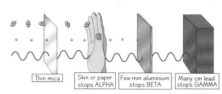

Thin mica | Skin or paper stops ALPHA | Few mm aluminium stops BETA | Many cm lead stops GAMMA

[1 mark for each material stopping correct radiation, total 4 marks]
2) For 0.6 Gy of alpha, the dose equivalent is $0.6 \times 20 = 12$ Sv [1 mark]. For 9 Gy of beta, the dose equivalent is $9 \times 1 = 9$ Sv [1 mark]. So, exposure to 0.6 Gy of alpha radiation would be more harmful than 9 Gy of beta radiation [1 mark].
3) $I \propto$ count rate $\propto \dfrac{1}{x^2}$ [1 mark]
   The GM tube is 4 times the original distance from the source.
   $\dfrac{1}{4^2} = \dfrac{1}{16}$, so the count rate at 40 cm will be 1/16ᵗʰ that at 10 cm [1 mark].
   The count rate at 40 cm will be $\dfrac{240}{16} = 15$ counts s⁻¹ [1 mark].

Page 87 — Nuclear Decay

1) a) $^{226}_{88}Ra \rightarrow \,^{222}_{86}Rn + \,^{4}_{2}\alpha$ [3 marks available — 1 mark for alpha particle, 1 mark each for proton and nucleon number of radon]
   b) $^{40}_{19}K \rightarrow \,^{40}_{20}Ca + \,^{0}_{-1}\beta$ [3 marks available — 1 mark for beta particle, 1 mark each for proton and nucleon number of calcium]
2) Mass defect $= (6.693 \times 10^{-27}) - (6.425 \times 10^{-27}) = 2.68 \times 10^{-28}$ kg [1 mark]. Using the equation $E = mc^2$ [1 mark], $E = (2.68 \times 10^{-28}) \times (3 \times 10^8)^2 = 2.41 \times 10^{-11}$ J [1 mark]

Page 89 — Exponential Law of Decay

1) a) Activity, A = measured activity – background activity $= 750 - 50 = 700$ Bq [1 mark]
   $A = \lambda N \Rightarrow 700 = 50\,000\,\lambda$ [1 mark] So $\lambda = 0.014$ s⁻¹ [1 mark]
   b) $T_{\frac{1}{2}} = \dfrac{\ln 2}{\lambda} = \dfrac{0.693}{0.014} = 49.5$ seconds
   [2 marks available — 1 mark for the half-life equation, 1 mark for the correct half-life]
   c) $N = N_0 e^{-\lambda t} = 50000 \times e^{-0.014 \times 300} = 750$
   [2 marks available — 1 mark for the decay equation, 1 mark for the number of atoms remaining after 300 seconds]

Page 91 — Binding Energy

1) a) There are 6 protons and 8 neutrons, so the mass of individual parts $= (6 \times 1.007276) + (8 \times 1.008665) = 14.112976$ u [1 mark]
   Mass of $^{14}_{6}C$ nucleus $= 13.999948$ u
   so, mass defect $= 14.112976 - 13.999948 = 0.113028$ u [1 mark]
   Converting this into kg gives mass defect $= 1.88 \times 10^{-28}$ kg [1 mark]
   b) $E = mc^2 = (1.88 \times 10^{-28}) \times (3 \times 10^8)^2 = 1.689 \times 10^{-11}$ J [1 mark]
   1 MeV $= 1.6 \times 10^{-13}$ J, so,
   Binding energy $(B) = \dfrac{1.689 \times 10^{-11}}{1.6 \times 10^{-13}} = 105.5$ MeV [1 mark]
2) a) Fusion [1 mark]
   b) Increase in binding energy per nucleon is about 0.86 MeV [1 mark]. There are 2 nucleons in ^{2}H, so the increase in binding energy is about 1.72 MeV — so about 1.7 MeV is released (ignoring the positron) [1 mark].

Page 93 — Nuclear Fission

1) a) $^{235}_{92}U + \,^{1}_{0}n \rightarrow \,^{90}_{36}Kr + \,^{144}_{56}Ba + 2\,^{1}_{0}n + $ energy
   [2 marks available — 1 mark for the proton number of Kr, 1 mark for the nucleon number of Ba]

Answers

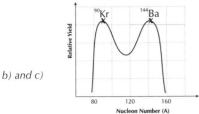

b) and c)

part b) [1 mark for 'm' shape of graph, 1 mark for labelling the x-axis to show that the peaks occur at about A = 90 and 140]
part c) [1 mark for placing Kr on the left-hand peak on the graph, 1 mark for placing Ba on the right-hand peak]

Page 95 — Nuclear Fusion

1) $T = \dfrac{E_k}{2 \times 10^{-23}} = \dfrac{1.1 \times 10^{-13}}{2 \times 10^{-23}} = 5.5 \times 10^9$ K   [1 mark]

2) a) mass defect = mass before − mass after
  = (2.013553 + 3.015501) − (4.001505 + 1.008665)
  = 0.018884 u
  [2 marks available — 1 mark for equation, 1 mark for value]

  b) 0.018884 × 931 = 17.6 MeV   [1 mark]

Page 97 — Classification of Particles

1) Proton, electron and electron antineutrino.   [1 mark]
  The electron and the electron antineutrino are leptons.   [1 mark]
  Leptons are not affected by the strong interaction, so the decay can't be due to the strong interaction.   [1 mark]

2) Mesons are hadrons but the muon is a lepton.   [1 mark]
  The muon is a fundamental particle but mesons are not.
  Mesons are built up from simpler particles.   [1 mark]
  Mesons feel the strong interaction but the muon does not. [1 mark]

Page 99 — Antiparticles

1) $e^+ + e^- \rightarrow \gamma + \gamma$ [1 mark]. This is called annihilation [1 mark].

2) The protons, neutrons and electrons which make up the iron atoms would need to annihilate with their antiparticles [1 mark].
  No antiparticles are available in the iron block [1 mark].

3) The creation of a particle of matter requires the creation of an antiparticle. In this case no antineutron has been produced [1 mark]. Also note that the total baryon number would have increased from 2 to 3 and that's not allowed.

Page 101 — Quarks

1) $\pi^- = d\bar{u}$ [1 mark]
  Charge of down quark = −1/3 unit.
  Charge of anti-up quark = −2/3 unit
  Total charge = −1 unit   [1 mark]

2) The weak interaction converts a down quark into an up quark plus an electron and an electron antineutrino.   [1 mark]
  The neutron (udd) becomes a proton (uud).   [1 mark]

3) The baryon number changes from 2 to 1 so baryon number is not conserved [1 mark]. The strangeness changes from 0 to 1 so strangeness is not conserved [1 mark].

Page 103 — Particle Accelerators

1) $f = \dfrac{Bq}{2\pi m} = \dfrac{20 \times 10^{-3} \times 1.6 \times 10^{-19}}{2\pi \times 1.67 \times 10^{-27}} = 0.305$ MHz
  [2 marks available — 1 mark for correct equation, 1 mark for correct value for f]

2) a) When a particle approaches the speed of light, its relativistic mass increases significantly. [1 mark]
  The particle's motion starts to get out of phase with the alternating p.d. [1 mark] which means the particle is no longer accelerated — putting a limit on the energy of a particle in the cyclotron. [1 mark]

  b) As the relativistic mass of a particle increases, the synchrotron increases the frequency of the alternating p.d. [1 mark]
  This keeps the particle accelerating in a circular path. [1 mark]

c) **Advantages**: [1 mark for any sensible advantage]
  The synchrotron is able to produce the highest energy particles.
  The synchrotron can accelerate two beams of particles in opposite directions at the same time.
  Disadvantages: [1 mark for any sensible disadvantage]
  The synchrotron cannot produce a continuous stream of particles.
  Synchrotrons generally need to be very large — this makes them expensive to build and maintain.

Page 105 — Detecting Particles

1) Charged particles follow curved tracks in a magnetic field [1 mark].
  +ve and −ve particle tracks curve in opposite directions [1 mark].
  Identify the direction of curvature for negative particles by looking for knock-on electrons OR apply Fleming's left-hand rule. [1 mark]

2) The proton and the positive pion give tracks but the neutron and the neutral pion do not.   [1 mark]

3)

[1 mark for two tracks going in opposite directions, 1 mark for not showing a track for the photon, 1 mark for tracks spiralling inwards.]
The tracks spiral inwards as the particles lose energy through ionisation.

Page 107 — Exchange Particles and Feynman Diagrams

1) The electrostatic force is due to the exchange of virtual photons that only exist for a very short time.   [1 mark]
  The force is due to the momentum gained or lost by the photons as they are emitted or absorbed by each proton.   [1 mark]

2)
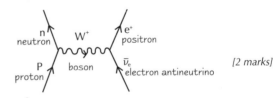

[2 marks]

This is a weak interaction.   [1 mark]

Section Ten — Astrophysics and Cosmology
Page 109 — The Solar System & Astronomical Distances

1) a) [1 mark each for 5 sensible points], e.g., planets are generally much bigger than comets. Planets are made out of rock and gas whereas comets are largely made of frozen substances like ice. Planets have almost circular orbits whereas comets have highly elliptical orbits. Comets have tails when they are close to the Sun. Comets can take millions of years to orbit the Sun; planets have much shorter periods.

  b) A comet has a tail because the energy from the Sun has started to melt it to produce a vapour trail [1 mark]. It always points away from the Sun as it is being 'blown' by radiation pressure [1 mark].

2) a) A light-year is the distance travelled by a photon of light in one year [1 mark].

  b) Seconds in a year = 365.25 × 24 × 60 × 60 = 3.16 × 10⁷ s [1 mark].
  Distance = c × time = 3.0 × 10⁸ × 3.15 × 10⁷ = 9.5 × 10¹⁵ m [1 mark].

  c) To see something, light must reach us. Light travels at a finite speed, so it takes time for that to happen [1 mark]. The further out we see, the further back in time we're looking. The Universe is ~15 billion years old so we can't see further than 15 billion light years. [1 mark].

Page 113 — Telescopes and Detectors

1) a)

Rays parallel to the principal axis of the lens converge to the principal focus [1 mark]. The focal length is the distance between the lens axis and the principal focus [1 mark].

Answers

b) $\dfrac{1}{u} + \dfrac{1}{v} = \dfrac{1}{f}$ [1 mark]. So $\dfrac{1}{0.2} + \dfrac{1}{v} = \dfrac{1}{0.15}$ [1 mark].
$\dfrac{1}{v} = \dfrac{1}{0.15} - \dfrac{1}{0.2} = \dfrac{5}{3} \Rightarrow v = 0.6\,m$ [1 mark].

c) Using lens equation: $\dfrac{1}{v} = \dfrac{1}{0.15} - \dfrac{1}{0.10} = -\dfrac{10}{3} \Rightarrow v = -0.3\,m$ [1 mark]

The sign of v is negative, indicating that the image is a virtual image on the same side of the lens as the object [1 mark].

2) a) Separation of lenses needs to be $f_o + f_e = 5.0 + 0.10 = 5.1\,m$ [1 mark].
b) Angular magnification = angle subtended by image at eye / angle subtended by object at unaided eye [1 mark].
$M = f_o/f_e$ [1 mark] $= 5.0 / 0.10 = 50$ [1 mark].

3) a) power $\propto$ diameter^2 [1 mark].

b) $\dfrac{power\ of\ Arecibo}{power\ of\ Jodrell\ Bank} = \dfrac{300^2}{76^2}$ [1 mark]. Ratio = 15.6:1 [1 mark].

4) a) When light strikes a pixel on a CCD, electrons [1 mark] are liberated from the silicon and are stored in a potential well [1 mark]. Once the exposure has been taken, electrons are shunted along the potential wells [1 mark] and emerge in sequence at the output, where they can be measured [1 mark].
b) CCDs have a quantum efficiency greater than 70% [1 mark] whereas photographic emulsion only has a 4% efficiency. So fewer photons are needed for an image and fainter objects can be detected [1 mark]. The output is in electronic form and can be processed digitally by computers, making it easier for the images to be enhanced [1 mark].

Page 115 — Spectra

1) The Sun has an 'atmosphere' that is cool enough for atoms to exist [1 mark]. The atoms absorb certain wavelengths of the continuous spectrum emitted from below [1 mark]. These wavelengths are then 'missing' (the atoms re-emit the light in all directions so less reaches the observer) leading to dark lines [1 mark].
2) a) The bright lines in the emission spectrum are in the same place as the dark lines in the absorption spectrum [1 mark].
b) A hot source (like a filament lamp) produces a continuous spectrum [1 mark]. This passes through a cold atomic gas of the element under observation [1 mark]. The resulting light is split up into its spectrum by a prism or a diffraction/reflection grating [1 mark].
c) Line emission spectra of all the elements are found and catalogued [1 mark]. The dark lines in the star's absorption spectrum are compared with these spectra. If the lines match up with a particular element then that element is in the star's atmosphere [1 mark].

Page 117 — Luminosity and Magnitude

1) The absolute magnitude is the apparent magnitude [1 mark] that the object would have if it were 10 parsecs [1 mark] away from Earth.
2) Distance to Sun in parsecs = $1/(2 \times 10^5) = 5 \times 10^{-6}$ pc [1 mark].
$m - M = 5 \lg (d/10)$ [1 mark] $\Rightarrow -27 - M = 5 \lg (5 \times 10^{-6}/10)$ [1 mark]
$\Rightarrow -27 - M = 5 \lg (5 \times 10^{-7}) \Rightarrow -27 - M = -31.5 \Rightarrow M = 4.5$ [1 mark].
3) a) Sirius is the brighter of the two [1 mark].
b) $m - M = 5 \lg (d/10)$ [1 mark] $\Rightarrow -0.72 - (-8.5) = 5 \lg (d/10)$ [1 mark]
$\Rightarrow 7.78 = 5 \lg (d/10) \Rightarrow \lg (d/10) = 1.556 \Rightarrow d/10 = 10^{1.556}$
So $d = 360$ pc [1 mark].

Page 119 — Stars as Black Bodies

1) a) $L = \sigma A T^4$ [1 mark] where L is the luminosity in W, σ is Stefan's constant (5.67×10^{-8} Wm^{-2}K^{-4}), A is the surface area in m^2 and T is the surface temperature in K [1 mark].
b) $3.9 \times 10^{26} = 5.67 \times 10^{-8} \times A \times 4000^4$ [1 mark], which gives $A = 2.7 \times 10^{19}$ m^2 [1 mark].
2) a) $\lambda_{max} \times T = 0.0029$ [1 mark].
So $T = 0.0029/(530 \times 10^{-9}) \approx 5500$ K [1 mark].
b) $L = \sigma A T^4$ [1 mark]. So $3.0 \times 10^{27} = 5.67 \times 10^{-8} \times A \times (5.5 \times 10^3)^4$, which gives $A = 5.8 \times 10^{19}$ m^2 [1 mark].

Page 121 — Spectral Classes and the H-R Diagram

1) a) A continuous spectrum is passed through hydrogen gas [1 mark] that is hot enough for electrons to be at the $n = 2$ excitation level [1 mark]. Electrons absorb photons of the correct energy (wavelength) to reach higher energy levels [1 mark]. Fewer photons with that energy reach the observer, leading to the absorption lines [1 mark].
b) To get strong Balmer lines, the majority of the electrons need to be at the $n = 2$ level [1 mark]. At low temperatures, few electrons have enough energy to be at the $n = 2$ level [1 mark]. At very high temperatures, most electrons will be at $n = 3$ or above, leading to weak Balmer lines [1 mark].
c) Spectral classes A [1 mark] and B [1 mark]
2) Molecules are only present in the lowest temperature stars [1 mark]. At higher temperatures molecules are broken up into atoms [1 mark].
3)

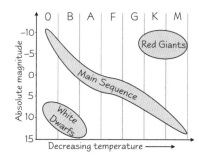

[5 marks maximum, 1 mark each for correctly labelled axes, 1 mark each for 'Main Sequence', 'White Dwarfs' and 'Red Giants'.]

Page 123 — Stellar Evolution

1) High mass stars spend much less time on the main sequence than low mass stars [1 mark]. As a red giant, low mass stars can only fuse hydrogen and/or helium but the highest mass stars can fuse nuclei up to iron [1 mark]. Lower mass stars eject their atmospheres to become white dwarfs [1 mark], but high mass stars explode in supernovae [1 mark] to become neutron stars [1 mark] or black holes [1 mark].
2) a) The Schwarzschild radius is the distance [1 mark] from the centre of a black hole to where the escape velocity is the speed of light [1 mark].

b) $R_s \approx \dfrac{2GM}{c^2} \approx \dfrac{2 \times 6.67 \times 10^{-11} \times 6 \times 10^{30}}{(3 \times 10^8)^2}$ [1 mark] ≈ 8.9 km [1 mark]

c) The formula is derived by using Newton's laws of motion and gravity. For intense gravitational fields, Einstein's theory of relativity is required for a more accurate answer [1 mark].

Page 125 — Galaxies and Quasars

1) $M = \dfrac{v^2 r}{G}$ [1 mark]. $r = 50\,000 \times 9.5 \times 10^{15} = 4.8 \times 10^{20}$ m [1 mark].

Substitute values: $M = \dfrac{500\,000^2 \times 4.8 \times 10^{20}}{6.67 \times 10^{-11}} = 1.8 \times 10^{42}$ kg [1 mark].

2) a) Their spectrum shows an enormous redshift [1 mark].
b) Intensity is proportional to 1/distance^2 [1 mark]. So if a quasar is 500 000 times further away than, but just as bright as, a star in the Milky Way it must be 500 000^2 times brighter than the star [1 mark].
c) A supermassive black hole [1 mark] surrounded by a doughnut-shaped mass of whirling gas [1 mark].

Page 127 — Models of the Universe

1) Kepler showed that the heliocentric model worked perfectly if the planets moved in ellipses [1 mark]. Galileo showed that the heavens were not perfect [1 mark], the phases of Venus were consistent with it orbiting the Sun [1 mark] and that Jupiter's moons showed that there were objects orbiting other bodies than the Earth [1 mark].
2) a) $v = H_0 d$ [1 mark] where v = recessional velocity, d = distance and H_0 = Hubble's constant [1 mark]
b) The Universe is expanding [1 mark], which suggests that it started as an infinitely hot, infinitely dense point [1 mark].
c) i) $H_0 = 50000 / 3.09 \times 10^{22} = 1.6 \times 10^{-18}$ [1 mark] s^{-1} [1 mark]
ii) $t = 1/H_0$ [1 mark] $= 6.2 \times 10^{17}$ s [1 mark] (20 billion years). So the observable Universe is a sphere of radius 20 billion ly [1 mark].

Answers

Page 129 — Evidence for the Hot Big Bang

1) a) $z \approx v/c$ [1 mark] so $v \approx 0.37 \times 3.0 \times 10^8 \approx 1.1 \times 10^8$ ms^{-1} [1 mark]

b) $d = v/H_0 \approx 1.1 \times 10^8 / 2.4 \times 10^{-18} = 4.6 \times 10^{25}$ m [1 mark]
= $4.6 \times 10^{25} / 9.5 \times 10^{15}$ ly = 4.9 billion ly [1 mark]

c) $z = v/c$ is only valid if $v << c$ — it isn't in this case [1 mark].

2) a) Object A is moving towards us [1 mark].

b) Object B is moving away from us [1 mark].

c) Object C is part of a binary star system (or is being orbited by a planet) [1 mark] with a period of two weeks [1 mark].

3) The cosmic background radiation is microwave radiation [1 mark] showing a perfect blackbody spectrum [1 mark] of a temperature of about 3 K [1 mark]. It is isotropic and homogeneous [1 mark]. It suggests that the ancient Universe was very hot, producing lots of electromagnetic radiation [1 mark] and that its expansion has stretched the radiation into the microwave region [1 mark].

Page 131 — Evolution of the Universe

1) a) When energy is converted into particles [1 mark], for every particle of matter created there is a corresponding antiparticle created [1 mark].

b) Matter and antimatter annihilate to form em radiation [1 mark]. If there were equal numbers then this annihilation would have been complete, leaving only radiation in the Universe [1 mark].

c) Theories predict that at high energies, there is a slight bias of matter over antimatter — which means that the annihilation left behind the small excess of matter that we observe today [1 mark].

d) Photons produced when matter and antimatter annihilated [1 mark].

2) A slight excess of matter was produced over antimatter [1 mark]. The matter and antimatter annihilated into photons [1 mark]. The remaining matter was in the form of quarks and leptons [1 mark]. As the Universe expanded and cooled the quarks combined to form particles like protons and neutrons [1 mark]. Some of the protons fused together to form helium [1 mark]. After about 300 000 years the Universe was cool enough for electrons to combine with the protons and helium nuclei to form atoms [1 mark]. The Universe became transparent since the photons could no longer interact with any free charges [1 mark]. There were fluctuations in the density of the Universe [1 mark] that allowed gravity to clump matter together [1 mark] condensing matter into stars and galaxies [1 mark].

Page 133 — Fate of the Universe

1) a) The average density of the Universe can be estimated using observations [1 mark]. This is close to the critical density needed to make the Universe flat [1 mark]. If both densities were projected back to the earliest Universe they agree incredibly closely [1 mark].

b) In a flat Universe, the angles of a triangle add up to 180° and parallel lines never meet [1 mark]. A flat Universe continues expanding forever [1 mark] but the expansion slows so that it approaches but never quite reaches a certain limiting size [1 mark].

2) a) $H_0 = 100000 / 3.09 \times 10^{22} = 3.2 \times 10^{-18}$ [1 mark] s^{-1} [1 mark]

$$\rho_0 = \frac{3H_0^2}{8\pi G} = \frac{3 \times \left(3.2 \times 10^{-18}\right)^2}{8\pi \times 6.67 \times 10^{-11}}$$ [1 mark] = 1.9×10^{-26} kgm^{-3} [1 mark]

b) Need 1.8×10^{-26} kg of matter for every cubic metre [1 mark]. So number of atoms needed = $1.9 \times 10^{-26}/1.7 \times 10^{-27} \approx 11$ [1 mark]

Section Eleven — Relativity
Page 135 — The Speed of Light & Relativity

1) a) The interference pattern would move/be shifted [1 mark].

b) The speed of light has the same value for all observers [1 mark]. It is impossible to detect absolute motion [1 mark].

2) a) An inertial frame is one in which Newton's 1st law is obeyed [1 mark], e.g. a train carriage moving at constant speed along a straight track (or any other relevant example) [1 mark].

b) The speed of light is unaffected by the motion of the observer [1 mark] or the motion of the light source [1 mark].

Page 137 — Special Relativity

1) $t = t_0\sqrt{1-\frac{v^2}{c^2}}$ [1 mark] and $t_0 = 20 \times 10^{-9}$ s [1 mark]

$$t = 20 \times 10^{-9}\sqrt{1 - \frac{0.995c^2}{c^2}} = 200 \text{ ns [1 mark]}$$

2) Your description must include:
A diagram or statement showing relative motion [1 mark].
An event of a specified duration in one reference frame [1 mark].
Measurement of the time interval by the other observer [1 mark].
Time interval for "external" observer greater than time interval for the "stationary" observer or equivalent [1 mark].

3) a) $m = m_0 \div \sqrt{1 - \frac{v^2}{c^2}} = 1.6 \times 10^{-27} \div \sqrt{1 - \frac{2.8^2}{3.0^2}} = 4.5 \times 10^{-27}$ kg [1 mark]

b) $E = mc^2 = 4.5 \times 10^{-27} \times (3 \times 10^8)^2 = 4.0 \times 10^{-10}$ J [1 mark]

Page 139 — General Relativity

1) Diagram of an accelerating frame [1 mark]
Diagram of a frame in a gravitational field [1 mark]
A statement saying that the situations are equivalent if the field strength is equal to the acceleration [1 mark]

2) Mercury's perihelion [1 mark], the point in the orbit where the planet is closest to the Sun, precesses [1 mark] at a very slow rate. The precession rate of Mercury's perihelion agrees with GR, but not Newton's theory of gravity [1 mark].

3) Observation: there is a change in the apparent position of a star [1 mark] whose light passes near the Sun [1 mark]. The apparent change in position lends support to the idea that light is deflected by the presence of massive objects as predicted by general relativity [1 mark].

Section Twelve — Medical Physics
Page 141 — Body Mechanics

1) 35% of body weight = $0.35 \times 700 = 245$ N [1 mark].
Compressive force = $245 \times \cos 50 = 157$ N [1 mark]
Shear force = $245 \times \sin 50 = 188$ N [1 mark]

2) In equilibrium, R = body weight. So R = 400 N [1 mark].
Taking moments about the point where C acts gives: 0.15R = 0.05T
So, T = 1200 N [1 mark].
The vertical forces must balance, so, R + T = C [1 mark]
giving C = 1600 N [1 mark].

Page 143 — The Physics of Vision

1) a) The distance will be the focal length of the lens [1 mark]
$v = f = 1$/power = 1/60 = 0.017 m [1 mark]

b) $1/u + 1/v = 1/f$, u = 0.3 m, v = 0.017 m [1 mark].
$1/f = 62.2$ D [1 mark]. So the extra power needed = 2.2 D [1 mark].

2) Far point is at infinity for a normally sighted eye [1 mark].
Image distance is 1/60 = 0.017 m [1 mark]. Max power = 70 D [1 mark], with a focal length of 1/70 = 0.014 m [1 mark].
Using the lens equation, $1/u + 1/v = 1/f$ gives $1/u + 60 = 70$ [1 mark].
Therefore $1/u = 10$. So near point = 10 cm [1 mark].

Page 145 — Response of the Eye

1) a) The brain receives signals from the retina in terms of relative strengths of red, green and blue [1 mark]. The relative intensities from a point on an image are seen as colour [1 mark]. Shining pure red, green and blue light sends an identical sort of signal to the brain [1 mark]. So any colour can be made by combining the primary colours [1 mark].

b) Nerve impulses take time to decay [1 mark] so an image stays on the retina for a short while. If the image changes within this time, the brain interprets it as smooth movement [1 mark]. TV pictures change quickly enough to fool the brain [1 mark].

2) a) Very sensitive to light intensity [1 mark]. Poor spatial resolution [1 mark]. Poor time resolution (so a greater persistence of vision) [1 mark]. Vision in greyscale rather than colour [1 mark].

b) Rod cells are initially saturated from too much light [1 mark]. Over about 40 minutes, the pigment returns due to the action of enzymes [1 mark]. Over this time, more and more rods become receptive to light, so the sensitivity of the retina increases [1 mark]. Rod cells share nerve fibres [1 mark], so the signal per nerve fibre is greatly amplified [1 mark].

Answers

Page 147 — Defects of Vision

1) a) Focal length of diverging lens needs to be –4 m [1 mark].
   Power = 1/f = –0.25 D [1 mark for value, 1 mark for negative sign]
   b) Lens equation 1/u + 1/v = 1/f [1 mark]. When u = 0.25 m [1 mark]
   v = –2 m ⇒ 1/f = 1/0.25 – 1/2 = 3.5. Power = +3.5 D [1 mark]
   c) Use bifocal lenses [1 mark] (or varifocal lenses).
2) a) Cylindrical lenses [1 mark].
   b)

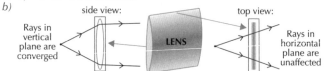

side view: top view:

Rays in vertical plane are converged   LENS   Rays in horizontal plane are unaffected

[2 marks maximum, 1 mark for each correctly drawn axis]

Page 149 — The Physics of Hearing

1) a) Using the principle of moments, F_1 × l must be equal to F_2 × $\frac{2}{3}l$,
   so $F_2 > F_1$ [1 mark].
   b) The oval window is smaller than the eardrum, so the intensity of the sound waves is increased [1 mark].

2) a) $v = \sqrt{\frac{E}{\rho}}$ [1 mark].

   so, E = 1000 × 1500² = 2.3 × 10⁹ Pa or Nm⁻² [1 mark].
   b) Attenuation is the reduction of intensity of a sound wave due to gradual absorption and dilution of its energy by the medium [1 mark]. Since attenuation is less in water than air, sound waves travel further underwater [1 mark].

Page 151 — Response of the Ear

1) a) 1 × 10⁻¹² Wm⁻² [1 mark]   b) 0 dB [1 mark]
2) a) Intensity is proportional to 1/distance², so doubling the distance quarters the intensity. New intensity = 0.25 MWm⁻² [1 mark].

   $IL = 10\log\left(\frac{I}{I_0}\right)$ and I_0 = 1 × 10⁻¹² Wm⁻² [1 mark]

   $IL = 10\log\left(\frac{0.25\times10^6}{1\times10^{-12}}\right) = 174$ dB [1 mark]

   b) The ear is most sensitive at about 3000 Hz, so the siren will sound as loud as possible [1 mark].

3) $IL = 10\log\left(\frac{I}{I_0}\right) \Rightarrow I = I_0 10^{\left(\frac{IL}{10}\right)}$ [1 mark] I_0 = 1 × 10⁻¹² Wm⁻² [1 mark]

   $I = 1\times10^{-12}\times10^{12} = 1$ Wm⁻² [1 mark]

Page 153 — The Heart

1) a) The potential difference between polarised and depolarised heart cells produces weak electrical signals at the surface of the body. This p.d. is plotted against time to give an ECG [1 mark].
   b) The right leg is too far away from the heart [1 mark].
   c) Dead skin cells and hairs are removed [1 mark]. Conductive gel is used to give a good electrical contact [1 mark].
2) a) A membrane is initially polarised so that one side is positively charged and the other is negatively charged [1 mark]. When the membrane is stimulated, it becomes permeable to sodium ions [1 mark]. The ions move through the membrane, depolarising the system [1 mark] and then polarising it the other way. In repolarisation, the membrane becomes impermeable to sodium, but very permeable to potassium ions [1 mark]. The potassium ions move through the membrane to reverse the polarisation [1 mark].
   b) The sinoatrial node produces ~70 electrical pulses a minute [1 mark]. These make the atria contract [1 mark]. They then pass to the atrioventricular node, which delays the pulses [1 mark] then passes them to the ventricles to make them contract shortly after [1 mark].

Page 155 — X-ray Imaging

1)

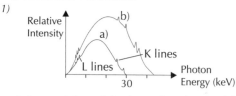

Relative Intensity   a)   b)   K lines   L lines   30   Photon Energy (keV)

a) See graph [1 mark for shape of graph, 1 mark for 30 keV maximum energy and 1 mark for correct labelling of line spectrum]
b) See graph [1 mark for higher intensity, 1 mark for higher maximum energy and 1 mark for a few extra lines in the line spectrum]
c) The photoelectric effect is the main mechanism for absorbing photons of this energy [1 mark]. The amount of absorption depends strongly on the atomic number [1 mark], so images of high contrast (between the different tissue types) can be obtained [1 mark].

2) $\mu = \frac{\ln 2}{x_\frac{1}{2}} = \frac{\ln 2}{3} = 0.23$ mm⁻¹ [1 mark], $I = I_0 e^{-\mu x} \Rightarrow \frac{I}{I_0} = e^{-\mu x}$

   So, 0.01 = e⁻⁰·²³ˣ ⇒ ln (0.01) = –0.23x [1 mark], x = 20mm [1 mark].

Page 157 — Ultrasound Imaging

1) a) $\alpha = \left(\frac{Z_{tissue} - Z_{air}}{Z_{tissue} + Z_{air}}\right)^2 = \left(\frac{1630-0.430}{1630+0.430}\right)^2$ [1 mark], α = 0.999 [1 mark]
   b) From part a), 0.1% enters body when no gel is used [1 mark].

   $\alpha = \left(\frac{Z_{tissue} - Z_{gel}}{Z_{tissue} + Z_{gel}}\right)^2 = \left(\frac{1630-1500}{1630+1500}\right)^2 = 0.002$ [1 mark], so 99.8% of

   the ultrasound is transmitted [1 mark]. Ratio is ~1000 [1 mark].
2) a) $Z = \rho v$ [1 mark], v = (1.63 × 10⁶)/(1.09 × 10³) = 1.50 kms⁻¹ [1 mark].
   b) Pulse from the far side of the head travels an extra 2d cm, where d is the diameter of the head [1 mark]. Time taken to travel this distance = 2.4 × 50 = 120 μs [1 mark]. Distance = speed × time, so 2d = 1500 × 120 × 10⁻⁶ = 0.18 m [1 mark]. So d = 9 cm [1 mark].

Page 159 — Magnetic Resonance Imaging

1)

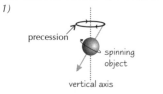

precession   spinning object   vertical axis

[2 marks available — one mark for the diagram showing the circular path of an object around the vertical axis, one mark for showing that the object must be spinning]
2) Protons will change alignment by either emitting or absorbing a specific amount of energy [1 mark].
   The amount of energy needed to do this is 2 μeV [1 mark].
3) There is hydrogen in all cells, so MRI can be used to examine any part of the body [1 mark]. Hydrogen only has one proton, so it can be made to emit specific EM frequencies that can be detected [1 mark].

Page 161 — Endoscopy

1) a) sin θ_c = n_2/n_1 = 1.30/1.35 [1 mark], θ = 74.4° [1 mark].
   b) The light will undergo total internal reflection [1 mark].
2) a) For transmitting an image of the area of interest in the body back to the observer [1 mark].
   b) You can have a greater number of very thin fibres in a bundle [1 mark]. This means that the image will have a higher resolution [1 mark].

Page 163 — Biological Effects of Radiation

1) a) The ionising energy of the particles released in 1000 s will be:
   1000 × 3 × 10¹⁰ × 8 × 10⁻¹³ = 24 J   [1 mark]
   Absorbed dose = ionising energy ÷ mass = 24 ÷ 70 = 0.34 Gy [1 mark]
   b) The dose equivalent of this absorbed dose is:
   H = 0.34 × 20 = 6.9 Sv   [1 mark]

Index

Index